Acknowledgements

The publishers would like to thank the following for permission to reproduce examination questions:

LEAG   London and East Anglian Group
MEG   Midland Examining Group
NEA   Northern Examining Association
SEG   Southern Examining Group

'New clothes for the Statue' is adapted, with permission, from *Education in Chemistry* 23, **5**, 1986.

Photographs

Armitage Shanks Ltd, p. 106; Arnica Management, p. 82; Clive Barda, p. 235; Barnaby's Picture Library, p. 143 (bottom right); BCIRA, p. 27; Biophoto, p. 34 (top); John Birdsall, p. 17 (bottom); Brewers' Society, p. 266; British Alcan Aluminium plc, p. 217; British Gas, p. 246; British Museum (National History), p. 137 (right); Building Research Establishment Crown Copyright photograph reproduced by permission of the Controller, H.M.S.O., p. 79 (bottom); J, Allan Cash Photo Library, pp. 26 (left), 79 (top), 81 (left), 99, 124, 144, 155, 178 (top left), 190, 204, 209 (bottom), 220, 225, 251, (bottom), 254, 274; Central Electricity Generating Board, pp. 95, 251 (top); Chubb Fire Security Limited, p. 159; De Beers, pp. 137 (left), 153 (bottom); De Beers Industrial Division, p. 153 (top); M.G. Duff Marine, p. 222; Education Service of the Plastics Industry, p. 262; Ever Ready Limited, p. 108 (right); Vivien Fifield, p. 47 (top); Friends of the Earth (C. Rose), p. 182; GeoScience Features, p. 209 (top); Griffen & George, p. 21; Adam Hart-Davis, p. 29; Trevor Hill, pp. 7, 17 (top), 28, 55, 85, 86, 108 (left), 149, 164, 189, 191, 221, 207, 231, 275; Holt Studios, p. 167 (left); ICCE (Philip Steele), p. 192; ICI Advanced Materials, p. 263; ICI Biological Products, p. 271; ICI Chemicals & Polymers, pp. 19, 160, 167 (right), 169, 170; Integrated Marketing Services, p. 26 (right); Colin Johnson, pp. 34 (bottom), 35 (top left and centre), 40, 87, 120, 121 (bottom), 130, 138, 180 (top left and right), 181, 185, 186, 226; P.V.A. Johnson, London Assay Office, p. 18; Lever Brothers Limited, p. 270; Longman Group, *Hunt & Sykes Chemistry*, p. 206; Osram-GEC Ltd, p. 143 (bottom left); Pilkington plc, p. 176 (bottom); Royal Mint, p. 203 (top); Science Photo Library, pp. 33, 36, 64, 81 (right), 89 (bottom), 143 (top); Seagram United Kingdom Limited, p. 268; S.P. Tyres (UK) Ltd, p. 178 (top right and bottom); Charles Tait, p. 252; Thames Water Authority, p. 188; Thermit Welding Limited, p. 228; United Glass Containers, p. 176 (top); Westland Helicopters Ltd, p. 136.

Illustrations by Taurus Graphics.

Contents

Introduction

Chemistry is as up-to-date as the clothes you wear, the food you eat, the air you breathe and the car you ride in. Almost very aspect of your daily lives is affected or controlled by Chemistry.

Chemists create new materials and test existing ones for purity or safety. They are involved in engineering, fuel technology, electronics, space travel and every other form of modern science.

This book sets out the important ground-work of Chemistry in a simple logical manner. If you are studying Chemistry as a separate subject, you could work through the book on your own, or with a teacher. If you are studying Integrated, Coordinated or Modular Science, then you will find the Chemistry topics here.

Each section has an introduction which shows how Chemistry is involved in everyday life and the Key Points at the end help you revise. There are quick questions and longer, examination-style questions too. More demanding questions are marked with a star.

A world without any manufactured materials at all would be a very difficult place in which to live, so it's just as well that we have Chemistry Now.

CHEMISTRY

Carshalton College

00011237

0007327

Learning Centre
Carshalton College
Nightingale Road
Carshalton, SM5 2EJ   SM5 2EJ
Tel: 020 8544 4344

Return on or before the last date stamped below.

2 6 MAR 2003

1 0 NOV 2005
1 4 FEB 2007

WITHDRAWN

E

6/90   43305

Oxford University Press 1989

Oxford University Press,
Walton Street, Oxford OX2 6DP

Oxford New York Toronto
Delhi Bombay Calcutta Madras Karachi
Petaling Jaya Singapore Hong Kong Tokyo
Nairobi Dar es Salaam Cape Town
Melbourne Auckland

and associated companies in
Beirut Berlin Ibadan Nicosia

Oxford is a trade mark of Oxford University Press

© Oxford University Press 1989

ISBN 0 19 914239 1

First published 1989

The cover photograph is by kind permission
of Science Photo Library/Chemical Design Ltd.

Printed in Great Britain by
Scotprint Ltd., Musselburgh

Working in a laboratory

1.1 *Safety*

Safety at home

The two substances shown in Figure 1 could be dangerous, but there are instructions on the side of them to tell you what not to do.

Why shouldn't young children be allowed to touch them?
Why could it be dangerous if you get them in your eyes or breathe their fumes in?
What might happen to the can if thrown into a fire or left in the sun?
Why shouldn't you remove the label until all the contents are used up?

Figure 1

Safety in the laboratory

Just like oven cleaner and lavatory cleaner, all bottles and jars of laboratory chemicals have labels on them saying what is in them, and explaining any dangers. The labels have a lot of information on them. Never leave any chemical — even a test tube — without a label on it. Look at Figure 2.

Figure 2
What is dangerous about this chemical? Why is it important to know how old it is?

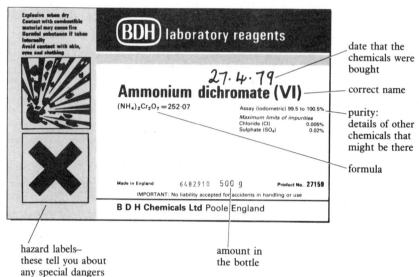

date that the chemicals were bought

correct name

purity: details of other chemicals that might be there

formula

hazard labels– these tell you about any special dangers

amount in the bottle

Hazard labels.

These explain themselves

What do these mean?

Laboratory rules

DO	DON'T
● keep bags and coat safely out of the way	● run in the laboratory
● tie up long hair and tuck in ties and loose clothing	● eat or drink in the laboratory
● wear goggles or safety glasses when doing experiments	● play with fire, electrical switches or chemicals
● keep your place tidy and wipe up spills of chemicals	● look down a test tube that is being heated, or point it at anyone
● ask if you are unsure about anything	

Using apparatus carefully

Here are a few things worth remembering.

Using a Bunsen burner.

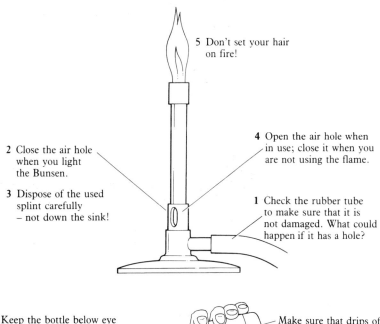

5 Don't set your hair on fire!

4 Open the air hole when in use; close it when you are not using the flame.

2 Close the air hole when you light the Bunsen.

3 Dispose of the used splint carefully – not down the sink!

1 Check the rubber tube to make sure that it is not damaged. What could happen if it has a hole?

Pouring liquid from a bottle

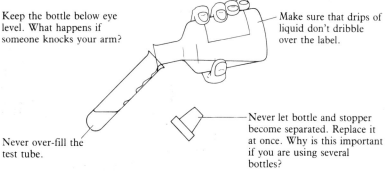

Keep the bottle below eye level. What happens if someone knocks your arm?

Make sure that drips of liquid don't dribble over the label.

Never over-fill the test tube.

Never let bottle and stopper become separated. Replace it at once. Why is this important if you are using several bottles?

Taking solid from a jar.

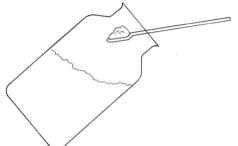

Use a clean spatula.

Don't over-load it or you might spill the chemical. Why could this matter?

Heating chemicals in a test tube.

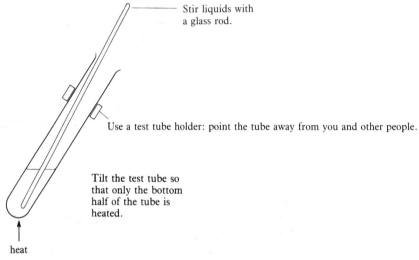

Stir liquids with a glass rod.

Use a test tube holder: point the tube away from you and other people.

Tilt the test tube so that only the bottom half of the tube is heated.

heat

Heating a flammable liquid.

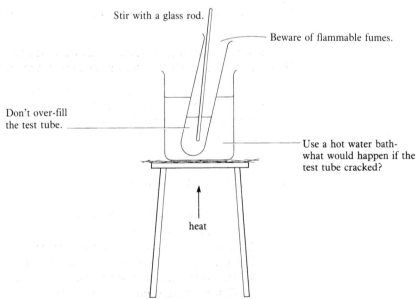

Stir with a glass rod.

Beware of flammable fumes.

Don't over-fill the test tube.

Use a hot water bath- what would happen if the test tube cracked?

heat

Key points

This section covered:

- safety in the laboratory
- the importance of well-labelled bottles
- taking solids and liquids from bottles correctly
- using a bunsen burner to heat solids and liquids

Quick questions

1 What safety warnings are there at (a) a petrol station (b) the swimming pool?

2 Ethanoic acid is a corrosive and flammable liquid. Draw the hazard labels that should be found on the bottle and write a list of warnings for its use.

1.2 *Experiments and records*

Experiments

Scientists ask questions.

What if I reduce the amount of sugar in this food? Will it taste different?
Do all colour paints last as well in the sun, or do some of them fade?
Why doesn't this fuel burn as well as others I have tested?
Does it contain a substance that will irritate skin? Can I safely market this new aftershave?
Will it work without putting this costly ingredient in? Can I cut down on costs?

To answer such questions, scientists do experiments and record results. From these, they make conclusions and try to answer their questions.

Scientists make notes about what they have done for two reasons. They need to remember all the details of their experiments as well as the results. Also, they must be able to tell other scientists about their experiments so that they can be copied or checked.

Designing an experiment

Before starting an experiment, ask yourself:

1 What do I want to find out?
2 How am I to set about the experiment?
3 What apparatus and materials will I need?
4 What measurements or observations will I make?
5 What safety precautions must I take?

Here is an example of an experiment.

Do oxides change mass when they are heated?

1 I need to find out if compounds ending in 'oxide' get heavier or lighter, or even stay the same mass when they are heated.

2 I must heat several oxides, weighing them before and after heating.

3 I shall need a balance to weigh the chemicals, a bunsen burner, test tubes, a test tube holder and a rack.

4 Obviously, I need to weigh the test tube and oxide before I start heating. I shall assume that the test tube doesn't change mass too.

After heating I shall weigh the test tube and oxide, and then heat it again to make sure that any chemical reaction has finished. I shall know when this has happened when the mass stops changing. (This is called 'heating to constant mass'.)

5 I shall wear a face mask all the time. I also need a metal test tube rack to put the test tubes in whilst they are cooling. I must be careful not to put very hot test tubes on to the balance pan.

Writing-up the experiment

The date when the experiment was done.

10.5.88

The heading, saying which experiment it was.

Do oxides change mass when they are heated?

The Aim says what the experiment was intended to do.

Aim

To find out if compounds with names ending in 'oxide' gain or lose mass, or stay the same when they are heated.

The Method should say exactly what was done, in the right order.

Method

First a test tube was weighed. After adding a small amount of the first oxide, the test tube and contents were weighed again. Using a medium bunsen flame, the oxide was heated for one minute and then cooled in a metal test tube rack. When cool enough to pick up, it was weighed again. It was then heated again and after cooling once more, weighed again. This was repeated until there was no further change in mass.

The experiment was repeated with the other oxides.

Results

Name of chemical	Mass of chemical before heating	Mass after first heating	Mass after second heating	Mass after third heating	Result
copper(II) oxide	30.15 g	30.15 g	30.15 g	30.15 g	same
copper(I) oxide	34.40 g	35.89 g	36.00 g	36.00 g	gain
red lead oxide	27.85 g	27.50 g	27.37 g	27.37 g	loss
lead(IV) oxide	47.90 g	46.35 g	46.30 g	46.30 g	loss
manganese(IV) oxide	33.55 g	33.55 g	33.55 g	33.55 g	same

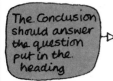
The Conclusion should answer the question put in the heading

Conclusions

Some oxides, such as copper(II) oxide and manganese(IV) oxide, do not change mass when heated. Some oxides, such as red lead and lead(IV) oxide, lose mass when heated. Some oxides, such as copper(I) oxide, gain mass when heated.

Making notes

When writing up your own notes, using a text book for the information,

DO:
- read the passage that you are using several times to get the ideas into your head
- decide what the main points of the passage are
- write the notes in your own words
- look up the meanings of words that you do not understand in a dictionary
- ask your teacher if you are unsure about anything you have read
- read through what you have written before handing it in.

DON'T:
- copy notes straight from the text book
- write down every single fact you can find whether it is relevant or not.

Key points

This section covered
- designing experiments
- writing-up experiments
- making notes from books.

Quick question

1 You have been asked to find out which of a series of chemicals dissolve in water, and which don't. Draw up a short plan of the way in which you would do the experiment.

Questions

1 These three students have just entered a chemistry laboratory.

Daryll   Emma   Gloria

For each student, explain why you would not allow them to do any experiments.

★ 2 This is a conversation between two pupils Sam and Jane doing a chemistry experiment. When you have read it:
   (a) write out the work sheet that you think the pupils might be using,
   (b) set out their results for them, in a table,
   (c) write a conclusion for them.

   Sam: Right, it says to get the substances and heat them.
   Jane: What substances?
   Sam: These three, in the beakers labelled A, B and C.
   Jane: What are they then? Oh, it says they are all carbonates.
   Sam: Yes A is copper(II) carbonate, that's $CuCO_3$, and C is lead(II) carbonate $PbCO_3$. The other one is nickel(II) carbonate.
   Jane: What's the formula for that?

Sam: I don't know. We shall have to look it up.

Jane: Now, it also says that we have to heat them in test tubes and weigh them after we have heated.

Sam: And before!

Jane: We can assume that the test tubes won't change in mass, so we don't have to weigh them separately.

Sam: The first one comes to one-five point four and its a green powder.

Jane: Look, it's going black as we heat it. Come on, let's weigh it again now.

Sam: No, we must wait until it cools down. Let's heat the others while we are waiting.

Jane: Well, B is dark green and it weighs fourteen point nine.

Sam: Good, it goes black too when it gets hot. Put it on one side to cool down. Now for C. It's white and two nine point seven.

Jane: This one goes yellow when it is hot. Write down all the descriptions because we are told to do that.

Sam: Right, now they are cold, let's weigh them again.

Jane: A comes to one-one exactly.

Sam: The next one is ten point five and C is two-five point three.

Jane: What does it say we have to do with the results?

Sam: Oh, something about seeing if carbonates gain or lose mass when they are heated.

2 *How matter is arranged*

2.1 *Pure substances and mixtures*

Rubbish

One of the best examples of a mixture is rubbish. It contains paper, food scraps, glass, aluminium and other metals, plastics and many other things. Each day, people in this country put 50 000 tonnes of rubbish into their dustbins. It may be rubbish to them, but to other people, it represents £900 000 000 worth of materials that can be recycled.

Waste paper can be turned back into clean paper, or used for packaging. It can also be 'fed' to special bacteria that convert it into animal food. **Scrap glass**, called 'cullet', is added to the chemicals used to make new glass. In this way, a lower temperature can be used and fuel is saved. Some estimates say that up to 12 000 000 litres of oil could be saved each year in this way. **Plastic** can be separated into useful chemicals, and mixtures of plastics can be made into black dustbin liners. Big squash bottles, made of Terylene, can be turned back into polyester fibre to make clothes. **Scrap metals** in rubbish are very valuable. 200 000 tonnes of aluminium are recycled each year saving £100 000 000 in imports of the materials needed to produce new **aluminium**. Expensive **tin** can be separated from **steel** in old food cans and steel from old cars is used in the production of new steel.

Figure 1
Look for this symbol on paper. It means that the paper has been made with recycled materials.

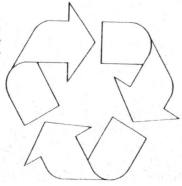

Figure 2
Both this shirt and bottle are made from polyester.

But separating these valuable materials can often be an expensive process. It is usually easiest and cheapest just to bury all the rubbish that comes from household rubbish bins because time and manpower would be needed to separate the recyclable things from the useless material. However, in some places, this cost has been offset by burning the remaining rubbish at power stations to produce heat for making electricity.

There are other ways of reclaiming substances before they are mixed up. Charities often collect waste paper and scrap aluminium from foil or cans. Bottle banks collect used bottles so that the glass can be recycled, and some people make a lot of money from collecting scrap metal.

Figure 3
Why do you think there are separate holes for brown, green and clear glass?

● What problems can you think of, if you wanted to set up a waste paper business? Start by thinking of a place to store the paper.

● Why don't we all just burn our rubbish in our gardens? Wouldn't that save someone collecting it?

What does 'pure' mean?

Rubbish is a **mixture** because it consists of iron, plastic, glass, paper and other things all mixed together. Substances like iron, aluminium, distilled water and copper sulphate crystals, however, are **pure substances**. If you look at them closely, they contain nothing but themselves. Iron contains only iron, distilled water contains only pure water and copper(II) sulphate crystals are just copper(II) sulphate and nothing use.

A pure substance is a single substance with nothing else mixed with it.

Gold

Gold is a soft metal, so pure gold rings and bracelets wear away quickly. To make them harder, the gold is mixed with small amounts of other metals such as copper, nickel, palladium and zinc. These mixtures are called **alloys**. The purity of gold is measured in **carats**. 24-carat gold is pure. 22-carat contains 2 parts of another metal. 18-carat gold contains even less gold and more of the other metals, and 14 and 9-carat golds are less and less pure. Whenever something is made of gold, it has to **assayed** (measured for purity). This is done at assay offices in for example, London, Bristol, and Birmingham. The assayers stamp a **hallmark** on the gold showing its maker, the year it was made, the office it was assayed at and its purity.

Figure 4
The symbols on this hallmark give information about the year and place of manufacture, as well as the purity. Can you decide which is which?

Air is a mixture. It contains several pure gases mixed together. When you take a deep breath, you breathe in oxygen, nitrogen, argon, carbon dioxide, water vapour and several other gases. When you drink a glass of water from the tap you are drinking another mixture. You are swallowing not only water, but the substances dissolved in it, like harmless chemicals from the rocks the water flowed over, air, and even chlorine, added at the water works.

> A mixture contains at least two substances mixed together. The substances can usually be separated somehow.

How can you tell if it's pure?

It is sometimes important to know whether a substance is pure, or whether there are impurities mixed with it. You can do this by measuring the **melting point**, if it is a solid, or the **boiling point**, if it is a liquid. Each pure substance has a particular melting point and boiling point which changes when an impurity is mixed in.

Figure 5
In winter, salt and grit are put on icy roads. The grit gives car tyres more grip. The salt lowers the melting point of the ice and makes it melt.

> Impurities make melting points lower and boiling points higher.

Different sorts of mixtures

Every pure substance is either a solid, a liquid, or a gas. Mixtures are combinations of pure substances. Here are some examples of some mixtures that you will know already.

- A mixture of metals is called an alloy. Fifty pence coins are made of copper and nickel mixed together.

- The sea is a solution. It contains salt dissolved in water.
- Fizzy drinks are a mixture of a gas (carbon dioxide) and water, along with flavouring and colouring.
- French salad-dressing is a mixture of two liquids, oil and vinegar, that do not mix. When shaken up, they form an emulsion, which slowly separates into layers again.
- Smoke is a mixture of tiny solid particles suspended in air.

Ways of separating mixtures

Filtering

Tea leaves are separated from tea with a strainer, wet lettuce is separated from water by a collander. Tiny particles of substances suspended in a liquid are separated using a **filter paper**.

Filtering dirty water

Dirty water contains tiny particles of grit and clay. When water is tipped through the filter paper, the particles are caught in the paper and clean water drips through. The particles are called the **residue** and the water is called the **filtrate**. The process is called **filtration**.

Figure 6
Filtering dirty water. The residue consists of mud. The filtrate is clear, but is it pure water?

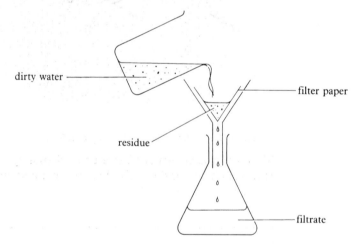

Centrifuging

Particles that are very tiny sometimes go through the small holes in a filter paper, or they clog up the holes and make the filtering process very slow. Mixtures of liquids and very tiny solid particles are called **suspensions**, because the particles do not sink to the bottom of the liquid but stay suspended. They can be separated using a **centrifuge**. Look at the photograph.

Figure 7
*As the centrifuge spins,
solid particles in the
suspension are pushed to
the bottom of the tubes,
leaving clear liquid
above.*

Dissolving, filtering and evaporating

This is a process that is often used in chemistry. Mixtures of solids can be separated if one of them dissolves in water and one does not. A good example is salt and sand.

Separating salt and sand

Water is added to the mixture and it is stirred. The **soluble** salt dissolves and forms a solution. The **insoluble** sand does not dissolve.

The mixture is tipped through a filter paper. The salt solution drips through. This is called the **filtrate**. The sand stays in the filter paper. It is called the **residue**.

The salt solution is gently heated. The water evaporates and leaves salt crystals.

Distillation

In the previous example, salt solution was evaporated to get crystals of salt. The water escaped into the air and was lost. If the solution had been boiled instead, the steam coming off could have been **condensed** back to pure water. This method of separation is called **distillation**.

Distilling sea water

Sea water is put into a distillation flask and boiled. Only pure water comes out of the solution and this goes into the **condenser** where it cools and changes back (condenses) to water.

Distillation is used in some countries to make drinking water from sea water. Ships also use distillation to purify sea water on long voyages.

Figure 8
When the liquid in the flask boils, only steam goes into the condenser. The thermometer will read 100°C. The condenser is kept cold by constantly flowing water from the tap.

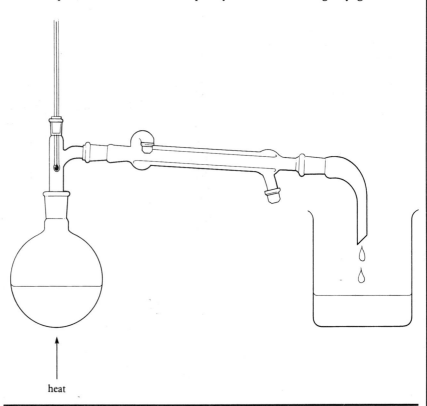

heat

Separating liquids that don't mix

Oil and water do not mix. They are said to be **immiscible** liquids. When they are shaken up together, they split into tiny drops and form an **emulsion**. An emulsion of oil and water quickly separates, with the denser water sinking to the bottom. Some emulsions like margarine stay mixed up and do not separate. This is because a chemical called an emulsifier is added too, and this helps hold the oil and water together.

Separating liquids that do mix

Distillation can be used to separate a mixture of liquids. This is an important process in the making of whisky, a mixture of ethanol and water. Mixtures of liquids can be separated by **fractional distillation**.

The fractional distillation of ethanol and water

The mixture of ethanol and water in the flask is heated so that it boils. Both ethanol vapour and steam go up the **fractionating column**.

Ethanol has a lower boiling point than water. Ethanol is said to be more **volatile** than water. In the fractionating column, steam condenses and drips back so only ethanol vapour gets to the top. Ethanol vapour changes to liquid ethanol in the condenser and drips into the collecting beaker.

Figure 9
The glass beads in the fractionating column separate the steam and ethanol vapour. Only ethanol vapour reaches the condenser. What happens to the steam?

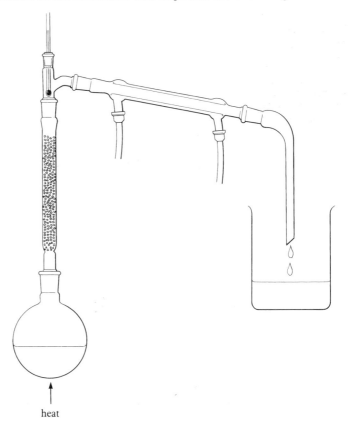

heat

Chromatography

This method is often used to separate biological material that has been obtained from plant or animal tissues. It can also be used to identify additives in foods such as flavouring and colourings.

Are the dyes in Smarties single colours?

Some of the colour is taken from a brown Smartie with a wet paintbrush. A small spot of colour is put on a piece of chromatography paper and allowed to dry. Look at the diagram.

The paper is clipped to a splint and balanced on top of a beaker. The end of the paper dips into a **solvent** (a mixture of butanol, water and ammonia). The solvent soaks up the paper and takes the spot of colour with it.

When the solvent has nearly reached the top, the paper is removed and dried. The spot of Smartie colour has separated into several other colours. Brown Smarties are coloured by a mixture of dyes. This method could be used to identify the colours used.

Samples of known food colours could be treated in the same way as the Smartie dye and compared.

Figure 10
Smartie chromatography.

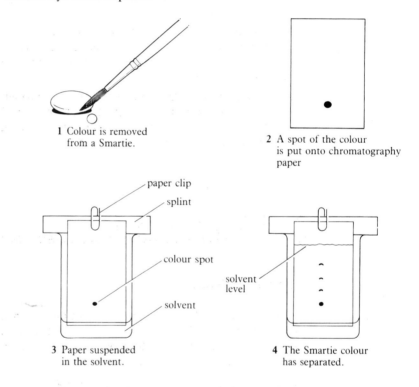

1 Colour is removed from a Smartie.

2 A spot of the colour is put onto chromatography paper

3 Paper suspended in the solvent.

4 The Smartie colour has separated.

Key points

- Many items of rubbish can be recycled.
- Pure substances have nothing mixed with them.
- Pure substances each have their own fixed melting points and boiling points.
- Mixtures called solutions can be separated by evaporation.

- Suspensions can be separated by filtration or centrifuging.
- Mixtures of miscible liquids can be separated by distillation.
- Dyes or colouring agents can be identified and separated by chromatography.

Quick questions

1 Name three substances that are thrown away in the dustbin that could be recycled. What could the recycled materials be made into?

2 Put the following substances into two lists: one for pure substances and one for mixtures.

   garden soil , copper, a two pence piece, river water, nitrogen, orange squash, custard, copper sulphate crystals, aluminium.

3 Fill in the missing words in this passage. Don't copy the passage out — just write down the words.

   Water taken from an . . . in the desert might . . . sand mixed with it. . . . separate pure water from . . . sand, the process of . . . could be used. On . . . voyages at sea, ships make . . . water from sea water . . . the process of distillation.

   If . . . salt is to be . . . from a mixture of . . . and small stones, the mixture . . . first added to water and . . . . The salt will dissolve, . . . not the stones. The . . . is then filtered. The residue . . . of the sand and . . . is . . . away. The filtrate is salt . . . . When it is evaporated, . . . of salt are left.

2.2 *Solids, liquids and gases*

The Kelvin Scale

0°C was chosen as the bottom of the Celsius scale, not because it was the lowest temperature that could be reached, but because water became frozen at that temperature. The freezing point of water is called a **fixed point**. Pure water always freezes at 0°C.

Temperatures can go much lower than that, however. A 19th century scientist called Lord Kelvin calculated the lowest possible temperature to be −273°C and called it **absolute zero**. His work gave us a new temperature scale called the **Kelvin scale**. Look at Figure 1.

Figure 1
The Kelvin scale — note the degree symbol is not used.

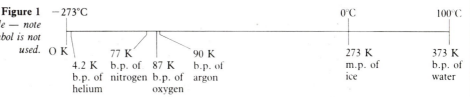

Many substances which are gases at normal temperature, condense and change into liquids at much lower temperatures. You can see the condensation points of oxygen, nitrogen and argon marked on Figure 1. Low temperature liquid gases are important. Liquid oxygen is used for rocket fuels and liquid nitrogen is used for deep freezing food.

Of all gases, helium has the lowest boiling point, at 4.2 K. When this liquid is cooled further to 2.19 K, strange things start to happen. Liquid helium becomes so runny that it creeps up the sides of the beaker and runs down the outside. The only way to keep it still is to freeze it and this happens only at 1 K and under very high pressure.

Solids, liquids and gases

Solids are often hard and strong: for example, steel and concrete. Bridges are often made of concrete and railway lines are made of steel. Most solids are rigid in shape and are often difficult to bend. However, when they are hot, they are easy to squeeze or stretch, although their overall volume does not change. At higher temperatures, they melt and change into liquids. Aluminium is a solid but when it is heated to 660°C, it becomes soft enough to be squeezed into different shapes. This process is called **extrusion**. Although the shape of the aluminium has changed, it still takes up the same amount of space — its volume has not changed.

Figure 2 *This substance fills the gap between the door and the wall. What would happen if it dissolved in water?*

Figure 3 *Hot aluminium can be squeezed, or extruded, into shape. Can you think of other objects made by this process?*

Iron is a solid. When it is heated to 1540°C, it **melts** and changes into a liquid. It can then be poured into moulds and cast.

Figure 4
Iron can be cast into different shapes when liquid. It quickly freezes into a solid and can then be removed from the mould. What other substances are cast like this?

Liquids take the shape of any container they are put into. Water is a liquid, but when it is cooled to 0°C, it **freezes** into solid ice and when heated to 100°C, it **boils** into gaseous steam. Brake fluid is a liquid. When a car driver puts a foot on the brake pedal, the brake fluid is pushed down a narrow tube which is connected to the brake. Like all liquids, brake fluid is very hard to squash (compress), so the brakes go on instantly.

Figure 5
What would happen to these brakes if brake fluid were easily compressed? Why it is so dangerous if air gets into the system?

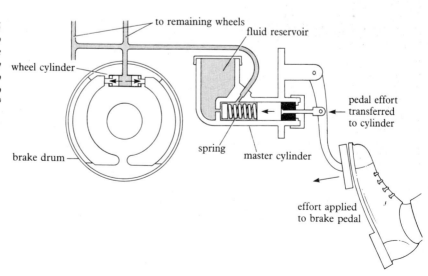

Gases are lighter than liquids and solids. They take up the shape of any container and can be compressed very easily. When cooled, they **condense** and change into liquids. Air is a mixture of gases. It takes up the shape of any container it is put into and can be compressed to very high pressures.

Figure 6
What does the gauge tell you about the air in the car tyre? Why do you think solid tyres are not used instead?

Steam is a gas. When it is cooled, it condenses and changes into water, and then freezes into ice. When heated, it just gets hotter and hotter.

The Kinetic Theory

Solids, liquids and gases are made of millions of tiny **particles** called **atoms** and **molecules**. These are covered in the next Section. In order to explain how the particles in solids, liquids and gases behave, scientists use a model called the **Kinetic Theory**. Kinetic means movement, so the theory provides a description of the way particles move in substances. In the Kinetic Theory, particles are thought of as small solid spheres, like miniature golf balls.

Figure 7
The arrangement of the particles in solids, liquids and gases is not the same. What are the two key differences?

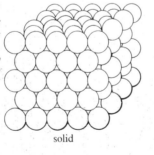

solid

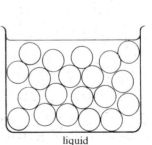

liquid

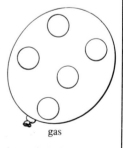

gas

In a solid, the particles are packed together very closely, in regular rows and layers. The particles cannot move around and can only vibrate in their places. This is why solids have definite shapes and why crystals have straight edges and flat surfaces. The closeness of the particles makes it difficult for solids to be compressed. The particles cannot be pushed any closer together. The closeness of the particles also makes many solids strong, and heavy.

In liquids, the particles are still very close together, but they have no fixed positions. They can move around freely. This is why liquids take up the shape of any container. The particles are still very close together, however, so liquids are very difficult to compress, just like solids.

In gases, the distance between the particles is very large. This is why gases can be compressed easily. The particles can be squashed closer together. Particles in gases are also moving around very quickly. Because of this, different gases mix easily. The fast moving particles quickly fit into each others' spaces.

Changing state

Solids, liquids and gases are called the three **states of matter**. When a solid melts or a liquid freezes, this is called a **change of state**.

Figure 8
What change of state is taking place here?

The Kinetic Theory can explain changes of state. When a solid is heated, the heat energy makes the particles vibrate. When the energy is great enough, the particles vibrate out of place and become jumbled up, just like a liquid. The solid has melted. The vibration overcomes the forces holding the particles together. If a liquid is cooled, the particles move more and more slowly until the same process happens in reverse.

When a liquid is heated, the heat energy makes the particles move faster and faster until they have enough energy to leave the liquid altogether and change into a gas.

Figure 9
The process of melting. What is happening to the particles?

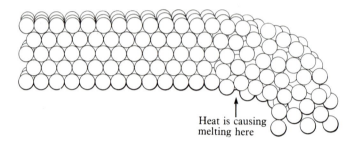

Heat is causing melting here

Cooling curves

If the temperature of a liquid is taken at intervals, as it cools down, the results can be plotted in a cooling curve. The point at which the liquid freezes can then be seen very easily. Each chemical freezes at its own particular temperature, so plotting a cooling curve enables you to find its melting point, or see how pure it is, or even to identify it.

Cooling curve apparatus

Wax is heated gently in a test tube until it melts. It is then stirred as it cools, and its temperature is taken each half minute until the wax is solid. The results are plotted on a graph.

Figure 10
The apparatus used to plot a cooling curve for wax. Why is it important to stir the wax before each temperature is taken?

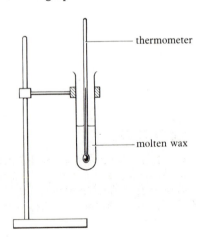

thermometer

molten wax

At the melting point (or in this case, the freezing point), the temperature remains constant all the time the liquid wax is changing into solid wax. The temperature does not fall again until all the wax is solid.

Figure 11
A cooling curve for wax. Melting and freezing take place at the same temperature. At what temperature did the wax melt?

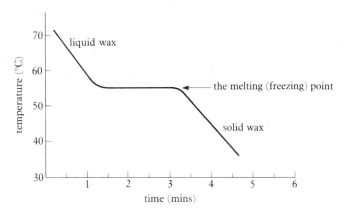

If the wax could be cooled from a very hot gas to a very cold solid, the complete cooling curve would look like Figure 12. Most substances have cooling curves like this, although the melting and boiling points will be different.

Figure 12
A complete cooling curve.

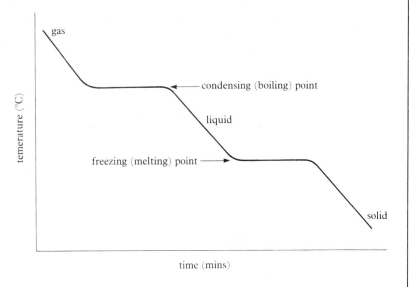

Key points

- Solids, liquids and gases all have easily distinguished properties.
- These properties can be explain with the Kinetic Theory.
- Melting/boiling and freezing/condensing are called changes of state.
- The melting point of a substance can be found by plotting a cooling curve.

Quick questions

1 Draw diagrams to show how the particles might be arranged in (a) a brick (b) water (c) oxygen.

2 A bath sponge has a regular shape and looks like a solid, but when you squeeze it, it squashes up to a fraction of its original size. How can you explain this?

3 Sketch the shape of the curve you might expect to get if ice was heated from absolute zero to 300 K, and its temperature was taken at regular intervals.

4 The three states of matter are solid, liquid and gas. In which state are the particles (a) moving fastest, (b) closest together?

2.3 Atoms and molecules

An early idea

Two and a half thousand years ago, a Greek philospher called Demokritos sat thinking about a piece of gold. In his mind, he divided it into two and then halved the two pieces into four, then eight and then sixteen, . . .

Eventually he decided that he could not divide the gold up for ever into smaller and smaller pieces. Eventually, he would come to a piece, so small, that it was impossible to split it further. So, he thought this particle must be one of the basic building blocks from which all matter was made. He called it atomos — indivisible. The atomic age began a long while ago.

Atoms

Atoms are very small. If about 2 million atoms were placed side by side, they would stretch just one millimetre. Each atom would be about 1/2 000 000 millimetres in diameter. On the tip of your little finger you could balance as many as 10 million atoms easily, and of course the finger itself, like everything that exists, is made of atoms.

> Everything is made of atoms. Atoms are the smallest particles of matter that can exist on their own.

Because they are so small, most atoms are impossible to see. Only the biggest atoms, like uranium atoms, have been seen as a blurred outline on a photograph taken through an **electron microscope**. Nevertheless, so many things happen that can only be explained by the idea of atoms, that scientists are sure they exist.

Figure 1
*Uranium atoms
magnified millions of
times. The small dots are
individual atoms, while
larger patches are clusters
of 2–20 atoms.*

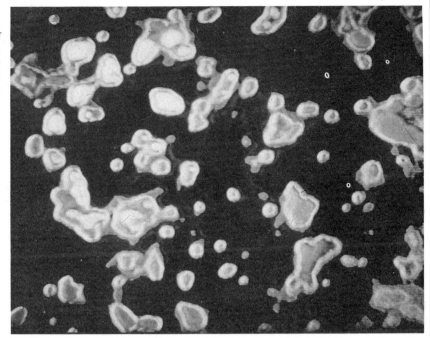

Molecules

Molecules are groups of atoms that are joined or
bonded together.

Water is made from hydrogen and oxygen atoms. It has the formula H_2O.
And yet water is a clear liquid and looks nothing like the gases hydrogen
and oxygen. This is because the hydrogen and oxygen atoms are **bonded**
together to make a molecule of water.

Figure 2
*This diagram represents
a water molecule. How
many atoms does it
contain?*

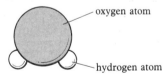

oxygen atom

hydrogen atom

Your body is a complicated structure of all sorts of substances, and yet it all
contain less than 30 different types of atom. These atoms join together
to make thousands of different molecules in blood, skin, flesh, bones, etc.

Most molecules are only just a little bigger than the atoms they are made
from, so they cannot be seen. Some molecules however, like the **DNA**
shown in the photograph contain tens of thousand of atoms, but they still
have to be magnified thousands of times to be seen with a powerful
microscope.

Figure 3
These chromosomes are made of DNA and thousands of atoms. Each cell of your body contains chromosomes.

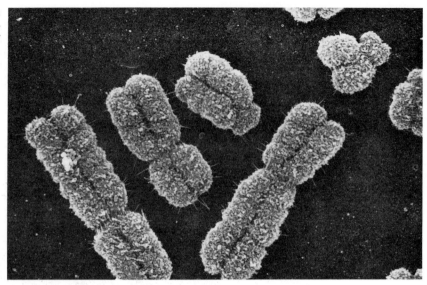

Reasons for believing in atoms and molecules

Many very simple things that we take for granted can only be explained if you believe matter is made of particles which scientists call atoms and molecules.

Crystals

The crystals of sodium chloride and alum in the photograph have flat surfaces and straight edges. This is because the atoms in the crystals always arrange themselves in regular rows and columns when the crystals are formed. They are always arranged in the same pattern too, so the crystals always have the same shape.

Figure 4
Crystals have straight edges and flat surfaces. Sodium chloride and alum always keep their own shape.

Figure 5
The atoms in a sodium chloride crystal always fit together in the same way. What shape is a sodium chloride crystal?

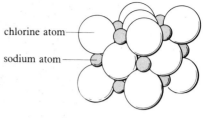

chlorine atom

sodium atom

Some crystals like **calcite** (a form of calcium carbonate) can be split by breaking between the rows of atoms with a sharp knife. This is called **cleaving**. Even very hard diamond crystals are cut and shaped in this way.

Figure 6
When this calcite crystal is split, it breaks between layers of atoms.

Dissolving

When a substance like sugar is added to water and stirred, the crystals dissolve and disappear. The molecules in the sugar crystals are arranged in a regular way, in rows and columns. The molecules of water are free to move about. Slowly, the sugar structure breaks up and the sugar molecules mix with the water molecules. The resulting mixture is called a **solution**.

Figure 7
When sugar dissolves, its molecules mix with water molecules. Which will there be the most of — water or sugar molecules?

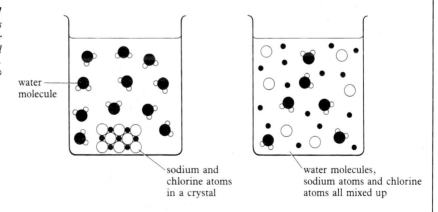

water molecule

sodium and chlorine atoms in a crystal

water molecules, sodium atoms and chlorine atoms all mixed up

Two liquids can mix and form solutions too. When ethanol is added to water, the molecules mix and a solution is made.

Diffusion

If someone fries bacon in the kitchen, everyone in the house soon knows about it! The smell tells you what is going on. The bacon is still in the pan, but the smell has drifted out into the air. This can be explained by saying that the molecules of 'bacon smell' are mixing with the air molecules and are slowly spreading out. This process of mixing gas molecules is called **diffusion**.

The photograph shows what happens when a gas jar of air is placed on top of a gas jar of brown nitrogen dioxide gas and the lids are removed. The gases start to mix and after a few minutes, have mixed completely. The process takes a little time because the molecules keep bumping into each other and slowing the process down. The diffusion happens faster if the gases are hot, because the molecules move faster.

Figure 8
This shows diffusion in action. It happens quite quickly. What does that tell you about the way molecules of gases move? How could you slow the process down?

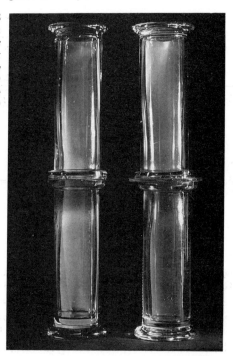

Brownian motion

In 1827, Robert Brown, a Scottish botanist, noticed a strange thing. While he was looking at some grains of pollen in water, under his microscope, he saw that they were moving about in a very jerky and erratic way. He thought that this might have something to do with the fact that they had come from a living plant, but about 50 years later, it was explained that the water molecules (far too small to be seen) were buffeting the pollen grains and making them move in this way. This **Brownian motion** can also be seen when tiny smoke particles are brightly illuminated and viewed through a powerful microscope. The smoke particles are bumped around by the air molecules. The air molecules cannot be seen, but the jerky smoke particles can.

Key points

- Atoms are the smallest particles of matter that can exist on their own.
- Molecules are groups of atoms joined together.
- Crystals have regular shapes because of the arrangement of their molecules.
- The processes of dissolving, diffusion and Brownian motion can be explained in terms of atoms and molecules.

Quick questions

1 Approximately how many atoms would stretch for 1 cm if placed side by side?

2 Molecules of carbon dioxide (the gas that you breathe out) are made of carbon atoms and oxygen atoms. Why can't you see black carbon and bubbles of oxygen when you breathe out a lungful of air?

3 Explain with diagrams, what happens to the molecules when (a) sugar crystals are added to hot water, (b) washing-up liquid is added to water, (c) the stopper is taken out of a perfume bottle.

2.4 *Elements and compounds*

Sir Humphrey Davy

Humphrey Davy was born in Penzance in Cornwall. He was apprenticed as a surgeon-apothecary, but as soon as he had finished his long training, he turned to his main interest, chemistry. Most of the things that he did in his early days would be considered dangerous or foolhardy now. In 1799, convinced that 'laughing gas' (dinitrogen oxide, N_2O) was breathable, he inhaled 18 litres of the gas over a period of 7 minutes. He made himself rather ill, but he demonstrated that laughing gas could be used as an anaesthetic.

He was made an assistant lecturer at the Royal Institution in London where he did many experiments that involved chemistry and electricity. In 1807, he passed electricity through molten sodium hydroxide and potassium hydroxide to make the new metals sodium and potassium. In the following year, he extracted boron, beryllium, barium, calcium, magnesium, lithium and aluminium, all elements that had not been made before.

Humphrey Davy travelled to many cities in Europe and carried on his research there. In Paris he showed that a newly discovered substance, iodine, was in fact a pure substance — an element. In Florence, he proved that diamond was also an element — just pure carbon, by burning diamonds with a giant magnifying glass.

- Do you think a scientist would breathe a large quantity of an unknown gas these days?

- Why hadn't electricity been used before this time in chemical experiments?

Elements

> Elements are pure substances that contain only one sort of atom.

A bar of pure gold contains only gold **atoms**. A lump of lead contains only lead atoms. The atoms in a piece of iron are all identical. So are the atoms in a piece of copper. But gold atoms, lead atoms, iron and copper atoms all have different sizes and different internal structures from each other.

Some non-metallic elements do not exist as individual atoms, but are made of groups of the same atom bonded together in **molecules**. The gases hydrogen, oxygen, nitrogen and chlorine are good examples. Other non-metals such as sulphur and phosphorus contain bigger molecules. Although they are made of molecules, these substances are still elements because the atoms in the molecules are identical.

Figure 1
These diagrams represent molecules of some well-known gases. The gases are elements because their molecules contain identical atoms.

 hydrogen   oxygen   nitrogen   chlorine

Figure 2
These elements have several atoms in their molecules.

phosphorus   sulphur

A few non-metals like oxygen, nitrogen, chlorine and argon are gases, but most elements are solids. There are only two elements which are liquids at ordinary temperatures. Mercury is a liquid metal and bromine is a liquid non-metal. A complete list of all the elements is given on page 282.

Symbols

The list of elements on page 282 shows the symbols of the elements. Each element has its own **symbol**. Often, the symbol is just the first one or two letters from its name.

Examples are:

hydrogen	H	oxygen	O	nitrogen	N
fluorine	F	carbon	C	calcium	Ca
chlorine	Cl	cadmium	Cd	caesium	Cs

Sometimes, the symbol does not match up with the name because it comes from the **Latin** or **Greek name** for the element.

Examples are:

iron, Fe (from ferrum); sodium, Na (from natrium); lead, Pb (from plumbum); and mercury, Hg (from hydragyrum).

Some elements get their names and symbols from **places**. Here are some:

copper, Cu (from the Latin for Cyprus); polonium, Po (after Poland); gallium, Ga (from the Latin for France); and californium, Cf (after California).

Other elements are named after **mythological characters** or heavenly bodies, such as: plutonium, Pu (after Pluto, god of the underworld) and helium, He (after Helios, the sun).

Some were named after **famous people** like nobelium, No (after Alfred Nobel) and einsteinium, Es (after Albert Einstein). Some are even named after **colours:** chlorine, Cl (Greek for green/yellow); platinum, Pt (Spanish for silver) and iodine, I (Greek for violet).

Compounds

> A compound is a pure substance containing different elements bonded together.

There are only 92 naturally occuring elements, but there are thousands of different compounds that can be made from them. Water and carbon dioxide are two simple compounds.

Molecules of steam are made when hydrogen and oxygen atoms react together.

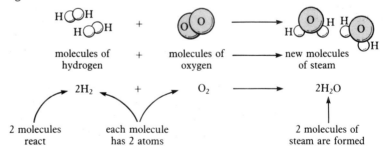

$$2H_2 \quad + \quad O_2 \quad \longrightarrow \quad 2H_2O$$

| 2 molecules react | each molecule has 2 atoms | 2 molecules of steam are formed |

These words and drawings can be represented by symbols and formulae.

| atom of carbon | + | molecule of oxygen | $\longrightarrow$ | molecule of carbon dioxide |
| C | + | O_2 | $\longrightarrow$ | CO_2 |

Carbon dioxide is made when carbon burns.

Changing elements into compounds

Sulphur and iron are elements. When they are shaken up together, they form a mixture, which can easily be separated with a magnet.

When the mixture is heated, the atoms of iron and the atoms of sulphur combine to make molecules of a new compound called iron(II) sulphide. The new compound is grey and brittle and completely different from the original mixture of yellow sulphur and black iron powder.

$$\text{iron} + \text{sulphur} \longrightarrow \text{iron(II) sulphide}$$
$$\text{Fe} + \text{S} \longrightarrow \text{FeS}$$

Figure 3
A mixture of iron and sulphur can be separated easily. Why isn't the sulphur attracted to the magnet?

Figure 4
The compound iron(II) sulphide is quite different to the mixture of iron and sulphur.

Understanding formulae

Every compound has a **formula**. The name and formula tells you what elements are present in the compound.

Here are some examples:

sodium chloride	NaCl	lithium chloride	LiCl
magnesium oxide	MgO	calcium oxide	CaO
iron(II) sulphide	FeS	zinc sulphide	ZnS

These names all end in 'ide'. The compounds are each made of just two elements only.

In the following compounds, the numbers tell you how many of each type of atom there are. They do not change the name.

aluminium oxide Al_2O_3 (2 aluminium atoms and three oxygen atoms)

sodium sulphide Na_2S (2 sodium atoms and 1 sulphur atom)

iron(III) chloride $FeCl_3$ (1 iron atom and 3 chlorine atoms)

These compounds all have names ending in 'ate':

copper(II) sulphate	$CuSO_4$	sodium sulphate	Na_2SO_4
calcium carbonate	$CaCO_3$	potassium carbonate	K_2CO_3
lead(II) nitrate	$Pb(NO_3)_2$	copper(II) nitrate	$Cu(NO_3)_2$

The **'ate'** shows the presence of a third element, oxygen.

Key points

- Pure substances containing only one sort of atom are called elements.
- Pure substances containing atoms of different elements joined together are called compounds.
- Each element has a symbol. Compounds have formulae made from the symbols of their elements.

Quick questions

1 Which of the following substances are elements?

sodium, calcium oxide, potassium, copper iodide, zinc, iron, aluminium sulphate, sulphur dioxide

2 Write down the symbols of the following elements:

(a) barium (b) phosphorus (c) silicon (d) argon (e) titanium (f) tungsten (g) fluorine (h) manganese.

3 Work out the names of these compounds:

(a) $MgCO_3$ (b) BaS (c) $Al_2(SO_4)_3$ (d) $Zn(NO_3)_2$ (e) $FeCl_2$ (f) SrO (g) KI

Questions

1 A sample of river water was purified using the apparatus below.

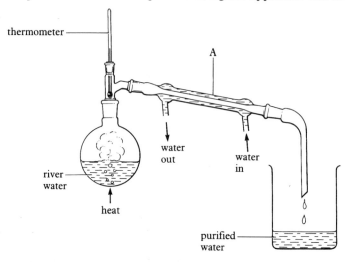

(a) What is this method of purification called?
(b) What is the purpose of the apparatus labelled A?
(c) What would the thermometer reading be when the river water is boiling?
(d) Why is this method of purification expensive when used on a large scale?
(e) How could you check that the purified water did not contain dissolved solids?
(Give your method and observations separately)

SEG

2 A young boy developed a rash every time he ate a packet of one particular brand of highly coloured sweets.
The local Food Inspector decided to use paper chromatography to test the sweets for the presence of banned food colourings.
She used ethanol as the solvent, and the chromatogram she obtained is drawn on the right.
(a) Describe how the dye colours could have been extracted from the sweets for use in the experiment.
(b) Explain why a pencil was used to draw the starting line rather than any other writing instrument.
(c) Which sweet(s) appear to contain only a single food colouring?
(d) (i) How is the colour in the green sweets made up?
(ii) Do the green sweets contain permitted colouring or banned colouring?
(e) The experiment did not give any useful information about the red sweets.
(i) Explain why not.
(ii) How might the experiment be modified to overcome this difficulty?

(f) (i) What decision should the Food Inspector make about the sale of these sweets?

(ii) Explain your answer to (i) as fully as you can.

MEG

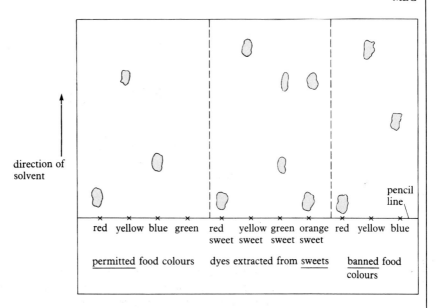

direction of solvent

pencil line

| red | yellow | blue | green | red sweet | yellow sweet | green sweet | orange sweet | red | yellow | blue |

permitted food colours   dyes extracted from sweets   banned food colours

3 The following is a list of the typical contents of a dustbin for a family of four in one week.

Aluminium	0.55 kg
Polythene/Plastic	0.25 kg
Waste food	4.50 kg
Iron	1.30 kg
Glass	1.75 kg
Paper	4.15 kg

Much of the waste in this dustbin can be changed back into useful metal, plastic, glass or paper products if it is processed correctly. First, some of the parts of the rubbish must be separated from the rest.

(a) (i) Give **one** method which could be used to separate objects made of iron from domestic rubbish.

(b) (i) Waste glass can be re-melted and used again. Give one problem which might be found when using empty bottles in this way.

(ii) Outline one **other** way in which the waste of glass containers can be avoided.

(c) (i) What percentage of the family's total waste is plastic?

(ii) Give **one** reason why plastic causes a serious pollution problem.

(iii) Plastic waste can be disposed of by burning it. Give one advantage and one drawback of this method.

(d) Explain why so many manufacturers use plastic packaging in spite of the pollution problems.

SEG

4 In this question, atoms of element X are shown as ○ and atoms of element Y are shown as ●
The boxes below represent different arrangement of elements X and Y.

A   B   C   D

Which boxes contain
(a) a mixture of X and Y?
(b) atoms only?
(c) molecules only?
(d) elements only?
(e) compounds only?

5 In a discussion on the use of rock salt, two pupils suggested that

ice melts more quickly if rock salt is added to it.

To test their prediction, they did the following experiment:
Step 1 Two funnels were filled with lumps of ice.
     2 Rock salt was added to one of the funnels.
     3 Water from the melting ice was collected in measuring cylinders.
     4 The volume of water was recorded every minute.

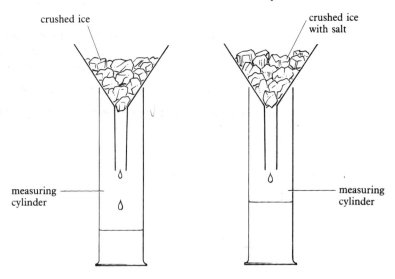

(a) State one precaution which should have been taken in Step 1 to make this a fair test.

(b) The volume of water collected from the ice *without* added salt is shown in the table below.

Time (minutes)	1	2	3	4	5	6
Volume of water collected (ml)	2	4	7	9	11	14

Draw a line graph to show how much water was collected as the ice melted and label this line A.

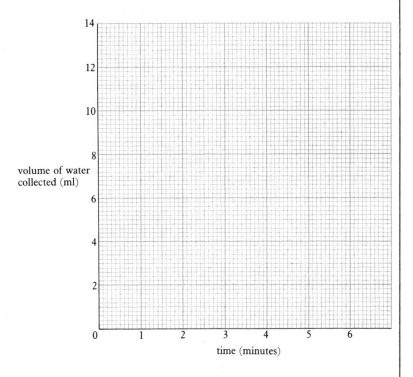

(c) The volume of water collected in the second measuring cylinder proved that the pupils' prediction was correct. Draw in another line on your graph to show this result and label this line B.

(d) Why does ice melt more quickly when rock salt is added?

(e) Gave one *disadvantage* of using salt on icy roads.

NEA

3 Inside the atom

3.1 *The structure of the atom*

John Dalton

John Dalton was born in a village called Eaglesfield in the Lake District in 1766. Later in that century, two laws had been discovered.

The Law of Conservation of Matter.

'Matter is not created or destroyed during a chemical reaction'.

The Law of Constant Composition.

'All pure samples of the same compound contain the same elements combined in the same proportions by mass'.

From these laws, and other work done by fellow scientists, John Dalton was able to make a list of rules about atoms. This is his list:

1 All matter is composed of atoms.
2 Atoms cannot be created or destroyed.
3 All the atoms in an element are identical.
4 The atoms of different elements are different.
5 When chemical reactions take place, atoms of different elements join together to make compound atoms. (We now call these molecules.)

John Dalton even made his own symbols for the elements that he knew. Look at the photograph.

Scientists now know much more about atoms. Atoms can be split and there are smaller particles inside. That is what the rest of this chapter is all about. However, John Dalton's rules about atoms are still very useful for describing in a simple way what atoms and molecules are, and for explaining how chemical reactions take place.

Figure 1
John Dalton used these symbols for the elements he knew. Would this system of symbols work with the number of elements we have today?

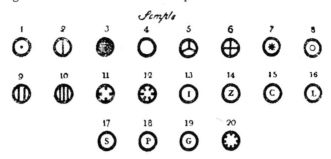

Inside the atom

By 1932, scientists had discovered that atoms contained three particles. The central part of the atom, called the **nucleus**, contained **protons** and **neutrons** and around the nucleus were smaller particles called **electrons**. These particles are very tiny. Their masses were first compared with that of a hydrogen atom. One hydrogen atom has a mass of one atomic mass unit. Here are some details about the particles.

Particle	Mass	Electrical charge
proton	1 atomic mass unit	positive
neutron	1 atomic mass unit	neutral
electron	1/1840 atomic mass unit	negative

You can see that protons and neutrons have equal mass. Electrons are much lighter.

Figure 2
Atoms contain protons, neutrons and electrons.

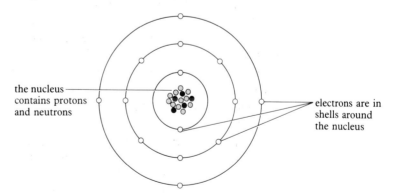

the nucleus contains protons and neutrons

electrons are in shells around the nucleus

Labelling atoms

Here are some important rules about atoms.

1 The number of protons in an atom is called the **atomic number** of that atom.

2 Each element has its own atoms and its own atomic number.

3 Atoms are neutral. The number of positive protons and the number of negative electrons in an atom are equal so that the charges cancel each other out.

4 The mass of an atom is found by adding the numbers of protons and neutrons together. The electrons are too small to affect the mass. This number of protons and neutrons is called the **mass number**.

atomic number = number of protons
mass number = number of protons + number of neutrons

It is often useful to give the mass number and atomic number of an element. This is how it can be written down.

mass number
              symbol of the element
atomic number

Here are some examples:

$$_1^1H \quad _6^{12}C \quad _7^{14}N \quad _9^{19}F \quad _{92}^{238}U$$

How many of each type of particle?

The number of protons, neutrons and electrons in an atom of an element can be worked out from its mass number and atomic number.

Here are some examples:

1      $_{26}^{56}Fe$

The atomic number of iron is 26.
This means it has 26 protons and 26 electrons in a neutral atom.
The mass number is 56
This means protons + neutrons = 56
So there are 56 − 26 = 30 neutrons.

2      $_{18}^{40}Ar$

Atomic number is 18, so 18 protons and 18 electrons.
Mass number is 40, so there are 40 − 18 = 22 neutrons.

Isotopes

Each element has its own atomic number. This is the number of protons in the nucleus of its atom. Different elements have different numbers of protons. However, some elements have two or more different atoms. These have the same number of protons, but different numbers of neutrons. These different atoms are called **isotopes**.

> Isotopes are different atoms of the same element.
> They have the same number of protons but different numbers of neutrons.
> Isotopes have the same atomic number but different mass numbers.

Here are some examples:

1 Hydrogen gas consists of two isotopes.

99.99% of the atoms are ^1_1H,
containing 1 proton and 1 electron.

0.01% of the atoms are ^2_1H,
containing 1 proton, 1 electron and 1 neutron.

^2_1H is sometimes called deuterium.

Hydrogen and deuterium are isotopes.

2 Chlorine gas consists of two isotopes.

75.40% of the atoms are $^{35}_{17}Cl$,
containing 17 protons, 17 electrons and 18 neutrons.

24.60% of the atoms are $^{37}_{17}Cl$,
containing 17 protons, 17 electrons and 20 neutrons.

Relative atomic mass

Each atom of an element has its own mass number. But, naturally occurring elements are mixtures of isotopes and so the masses must be averages of the different masses. Modern measurements of masses of atoms are made on an instrument called a **mass spectrometer.** This compares the average mass of the isotopes for an element with $^{12}_6C$. $^{12}_6C$ is used as a standard. These averaged, compared masses are called **relative atomic masses**. The value will be nearest in size to the mass of the most abundant isotope.

Relative atom mass (r.a.m.) has no units. It is just a number to compare the masses of elements.

Look at these examples:

1 Hydrogen is made of: 99.99% of ^1_1H ⎫
                                                ⎬ the r.a.m. is 1.008
                                 0.01% of ^2_1H ⎭

It is very near to 1 because most of the hydrogen atoms have a mass of 1. Only a tiny number have a mass of 2.

2 Chlorine is made of: 75.40% of $^{35}_{17}Cl$ ⎫
                                                     ⎬ the r.a.m. is 35.5
                                 24.60% of $^{37}_{17}Cl$ ⎭

This is near to 35 because most of the atoms are higher ones. Only $\frac{1}{4}$ have a mass of 37.

There is a list of relative atomic masses on page 282. You can see that most of them are not whole numbers, because they are mixtures of isotopes.

Electrons

The electrons in an atom are found in shells around the nucleus. The shells are not hard like an eggshell, but shapes that the electrons occupy. Each shell can only contain a certain number of electrons. When it is full, the next shell is used.

> The first shell is full when it contains 2 electrons.
> The second shell is full when it contains 8 electrons.
> When the third shell contains 8 electrons, the fourth shell starts to fill up.

Electron configurations

When electrons fill up the shells of an element, they always fill the shells nearest the nucleus first.

Here are some examples:

1 Carbon has the atomic number 6. An atom of carbon has 6 electrons. They are arranged like this:

first shell   2 electrons (it is now full)
second shell   4 electrons

The electron configuration is 2.4.

2 Chlorine has the atomic number 17. An atom of chlorine has 17 electrons. They are arranged like this:

first shell   2 electrons (it is now full)
second shell   8 electrons (it is now full)
third shell   7 electrons

The electron configuration of chlorine is 2.8.7.

3 Calcium is element number 20. An atom of calcium has 20 electrons. They are arranged like this:

first shell   2 electrons (it is now full)
second shell   8 electrons (it is now full)
third shell   8 electrons (the next shell now fills)
fourth shell   2 electrons.

The electron configuration of calcium is 2.8.8.2.

Key points

● Atoms consist of a nucleus containing protons and neutrons, surrounded by electrons.

- The number of protons in an atom is called its atomic number.
- The number of protons, plus the number of neutrons in an atom, is called its mass number.
- Atoms with the same atomic number, but with different mass numbers, are called isotopes.
- Most elements occur as a mixture of isotopes. The average mass of these isotopes is called the relative atomic mass.
- Electrons are arranged in shells around the nucleus of the atom.

Quick questions

1 For each of the following elements, write down its atomic number, mass number and numbers of protons, neutrons and electrons.

(a) $^{39}_{19}K$ (b) $^{40}_{20}Ca$ (c) $^{55}_{25}Mn$ (d) $^{127}_{53}I$

2 Write down the symbols, showing mass number and atomic number, for the elements with the following numbers of particles:

	Protons	Neutrons	Electrons
(a)	3	4	3
(b)	8	8	8
(c)	15	16	15
(d)	16	16	16
(e)	22	26	22

3 An element contains 50% of an isotope with mass number 48, and 50% of an isotope with mass number 50. What is the relative atomic mass of the element?

4 Write down the electron configurations of the following elements:

(a) fluorine (b) sodium (c) aluminium (d) sulphur (e) argon.

3.2 The periodic table

Dmitri Mendeleef

At the start of the 19th century only 33 elements were known. 60 years later, nearly double that number had been discovered and named. Chemists listed each element by its properties (its description and reactions), and by 1850, they were weighing the elements and comparing their masses with that of the lightest element known, hydrogen. They called these weights **atomic weights**. Atomic weights are now called **relative atomic masses**.

In 1869, a Russian chemist called **Dmitri Mendeleef** published a 'Periodic Table' in which he listed all the elements in order of their 'atomic weight', and with elements of similar properties underneath each other in the table. The first part of the table looks very much like the one we use today.

Figure 1
*The first periodic table,
written by Mendeleef.
Compare it with the
modern one. Is it
similar? Can you see any
elements that were not
discovered in Mendeleef's
time?*

	Group I	Group II	Group III	Group IV	Group V	Group VI	Group VII	Group VIII
1	H							
2	Li	Be	B	C	N	O	F	
3	Na	Mg	Al	Si	P	S	Cl	
4	K	Ca	?	Ti	V	Cr	Mu	Fe, Co, Ni, Cu
5	?	Zn	?	?	As	Se	Br	

He left gaps for elements that had not yet been discovered but which he felt must exist. Many people thought his ideas untrue, but it was not long before gallium was discovered in 1875 and slipped neatly into one of the spaces Mendeleef had left, and his table of elements became accepted.

The modern periodic table

The idea that electrons are contained in shells around the nucleus of the atom was put forward by a Danish chemist called Neils Bohr. He also built up a periodic table by putting the elements in the order of their atomic number (number of electrons). Every time he came to a full shell, he started another row.

The first shell can have a maximum of 2 electrons.

atomic number:	1	2
element:	H	He

The first shell was now full, so he started a new row.

atomic number:	3	4	5	6	7	8	9	10
element:	Li	Be	B	C	N	O	F	Ne

The second shell was now full, so he started a new row.

atomic number:	11	12	13	14	15	16	17	18
element:	Na	Mg	Al	Si	P	S	Cl	Ar

This shell now has 8 electrons, so he started another row.

atomic number:	19	20
element:	K	Ca

and so on.

Neils Bohr put all of the known elements into order like this, according to how many electrons they had in each shell, and this is what the finished periodic table looked like.

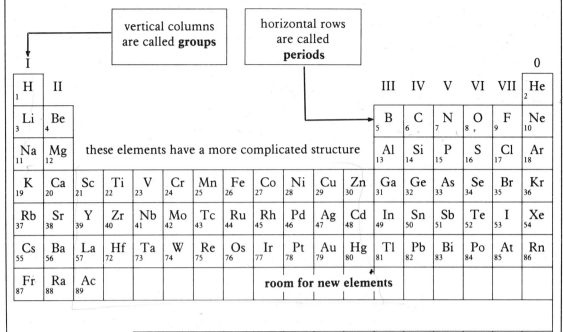

You can see that after calcium, the arrangement of the electrons becomes a little more complicated.

There are three important things that the table can tell you.

1 Horizontal rows of elements are called **periods**. Elements in the same period have the same number of shells of electrons.

2 Vertical columns of elements are called **groups**. The groups are numbered Group I to Group VII, and Group 0. Elements in the same group have the same number of electrons in their outer shell.

3 Elements in the same group have very similar properties because they have the same number of electrons in their outside shell.

Using the periodic table

The electron configuration of the first 20 elements can be obtained from the periodic table. After that, the structure becomes too complicated to do this.

Here are some examples:

1 The element aluminium is in the third period.
   This means that it has three shells of electrons.

Aluminium is in group III of the periodic table.
This means that it has three electrons in its outer shell.

The first and second shell are full.
The electron configuration of aluminium is 2.8.3.

2 The element potassium is in the fourth period.
It has four shells of electrons.

It is in group I.
It has one electron in its outer shell.

The other shells are full.
The electron configuration is 2.8.8.1.

Key points

- the modern periodic table is arranged in rows and columns called periods and groups.
- elements in the same period have the same number of shells of electrons.
- elements in the same group have the same number of outside shell electrons.
- the electron configuration of an element is the number of electrons that it has in each shell.

Quick questions

1 Which of the following elements (shown by their electron configurations) are in the same (i) period (ii) group of the periodic table?

(a) 2.1 (b) 2.8.3 (c) 2.8 (d) 2. (e) 2.8.8.1 (f) 2.8.5 (g) 2.8.8.2
(h) 2.8.8 (i) 2.5 (j) 2.7

3.3 Bonding

Silicones

Some people splash about a lot in the bath. To stop the water running down the walls and getting between the bath and the wall, the crack has to be filled and sealed. Plaster would simply get soggy and paint would crack every time the bath moved slightly. A sealing agent is used that is water repellent, flexible and attractive to look at. Figure 1 shows a similar substance used to seal around doors and windows. It is squeezed from a tube as a gluey substance and sets within an hour of being exposed to the air. It is completely water repellent.

Anoraks are designed to keep the wind and rain out. Cotton or even nylon are not completely water-proof so they have to be coated with a chemical that resists and repels water. The photo shows a treated material.

Figure 1 *This substance fills the gap between the door and the wall. What would happen if it dissolved in water?*

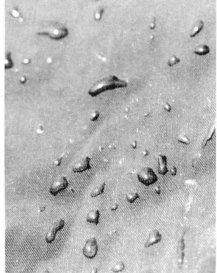

Figure 2 *The chemical on this material stops the water soaking in.*

These two substances have one thing in common. They contain compounds called **silicones**. Silicones have a structure which makes it difficult for them to mix with water, and so water doesn't soak into them. Silicones have a **covalent** structure.

- Do compounds like sodium chloride (salt) and copper(II) sulphate mix easily with water? Would you call them covalent substances?

Why elements react

An electric light bulb contains a thin filament of **tungsten** wire inside a glass bulb filled with **argon** gas. When electricity flows through the wire, it glows white hot and produces a lot of light. But the tungsten wire does not burn or change into anything. This is because argon is a very unreactive element.

Look at the electron configuration of argon:

Ar 2.8.8 All three shells are filled with their
maximum number of electrons.

Argon is very **unreactive** because its outside shell is **full**.

Now look at the electron configurations of two other elements in Group 0 of the periodic table:

He 2
Ne 2.8 Again, all the shells are full with their
maximum number of electrons.

Helium and neon are very unreactive elements because they also have full outside shells.

Elements that have full shells of electrons are stable
and very unreactive.

Elements that do not have full shells of electrons
react in order to reach that state.

Sodium and chlorine are very reactive elements. When a piece of sodium
is heated and allowed to burn in chlorine gas, there is a bright yellow flame
and white sodium chloride is formed.

sodium + chlorine ⟶ sodium chloride

Look at the electron configurations of sodium and chlorine:

sodium 2.8.1 A sodium atom needs to **gain** 7 more electrons or to **lose**
1 electron to make its outside shell a full one.

chlorine 2.8.7 A chlorine atom needs to **gain** 1 electron or to **lose** 7
electrons to make its outside shell full.

This is what happens when a sodium atom reacts with a chlorine atom.
The diagrams show the number of electrons each atom has in its **outer
shell**:

| sodium **loses** its electron | chlorine **takes** the electron | this sodium atom has **lost** an electron | this chlorine atom has **gained** an electron |

Atoms that have lost or gained electrons are called
ions.

Because the sodium atom has **lost** one negative electron, it becomes a
positive ion.
Because the chlorine atom has **gained** one electron, it gains a negative
charge and becomes a **negative ion**.

$$Na + Cl \longrightarrow Na^+ \quad Cl^-$$
atoms   ions

Sodium chloride is made of sodium ions and chloride ions.

Here are some more examples:

1 Magnesium burns in chlorine to form magnesium chloride.

magnesium + chlorine ⟶ magnesium chloride

Magnesium has the electron configuration 2.8.2.
It needs to **gain** six electrons, or **lose** two, to get a full outside shell.

Chlorine has the electron configuration 2.8.7.
It needs to **lose** seven electrons or **gain** one.

The diagrams show the electrons in the **outer shell** of each atom.

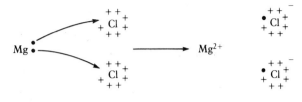

The magnesium atom **loses** 2 electrons.	The chlorine atoms **each gain** 1 electron.	The magnesium and chlorine atoms now have **full shells** and have changed into ions

The magnesium atom has lost two electrons and has gained two positive charges.
The chlorine atoms have each gained one electron and each therefore gains a negative charge.
Magnesium chloride is made of magnesium and chloride ions.

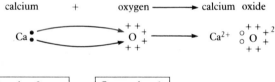

Calcium has 2 electrons in its outer shell. It **loses** them.	Oxygen has 6 electrons in its outer shell. It **gains** the 2 electrons from the calcium.	The calcium and oxygen atoms now have full shells and have changed into ions.

2 Can you follow what is happening when calcuim reacts with oxygen?

Calcium has lost two electrons and has gained two positive charges.
Oxygen has gained two electrons and has gained two negative charges.
Calcium oxide is made of calcium ions and oxide ions.

When elements react, they do so in order to get a full outer shell of electrons. When they have done this, they become stable and unreactive and have changed into ions.

Elements that lose electrons change into positive ions.
Elements that gain electrons change into negative ions.

This type of reacting, or bonding, is called **ionic bonding**. Ionic compounds contain ions. Because ionic compounds are made by the loss and gain of electrons, they are sometimes called **electrovalent compounds**.

Ions and the periodic table

Atoms can react by losing or gaining electrons and changing into ions. If they have a small number of electrons in their outside shell, they lose them and change into positive ions. If their outside shell is nearly full, they take in electrons to fill it up. This information can be found in the periodic table.

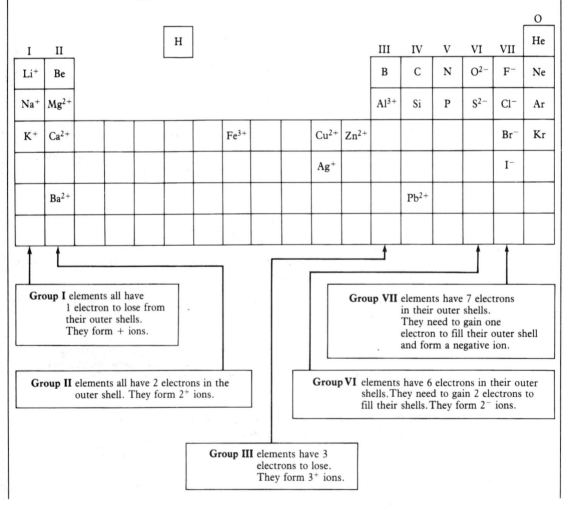

Group I elements all have 1 electron to lose from their outer shells. They form + ions.

Group VII elements have 7 electrons in their outer shells. They need to gain one electron to fill their outer shell and form a negative ion.

Group II elements all have 2 electrons in the outer shell. They form 2+ ions.

Group VI elements have 6 electrons in their outer shells. They need to gain 2 electrons to fill their shells. They form 2− ions.

Group III elements have 3 electrons to lose. They form 3+ ions.

Not all elements form ions, however. Group 0 elements do not react and form ions because they have full shells already. Other elements do not form ions, but bond in another way without forming ions.

Covalent bonding

Elements that bond without forming ions do so by sharing electrons to fill their outer shells. This is called **covalent bonding** or **covalency**.

Look at the examples:

Hydrogen chloride gas is formed when hydrogen atoms bond with chlorine atoms.
Hydrogen has the electron configuration 1.
Chlorine has the electron configuration 2.8.7.
Hydrogen needs one more electron to fill its outer shell. (Remember the first shell contains only two electrons.)
Chlorine needs one more electron to fill its outer shell.
Both elements do this by **sharing**:

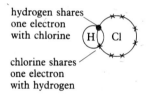

hydrogen shares
one electron
with chlorine

chlorine shares
one electron
with hydrogen

Hydrogen chloride is not made of ions. Atoms bonding in this way, without forming ions make uncharged **molecules**.

Water is made when two hydrogen atoms are bonded to an oxygen atom. Each atom gets a full outer shell of electrons by sharing.

Water is made of molecules. The oxygen atom shares 2 electrons, and each hydrogen atom shares 1 electron.

Methane is made by one carbon atom sharing electrons with four hydrogen atoms. Methane is made of molecules.

By sharing electrons, the carbon and hydrogen atoms fill their outer shells.

Ammonia is made when one nitrogen atom bonds with three hydrogen atoms.

> Covalent compounds are made when atoms fill up their outer shells of electrons by sharing electrons with other atoms. Covalent compounds are made of molecules.

Differences between ionic and covalent compounds

Ionic and covalent compounds behave in different ways because of the way that their atoms are bonded together.

Melting points

Ionic compounds generally have higher melting points and boiling points than covalent compounds. Here are some examples:

substance	melting point °C	boiling point °C
sodium chloride	801	1465
magnesium chloride	714	1418
calcium oxide	2600	2850
hydrogen chloride	−114	−85
water	0	100
methane	−182	−161
ammonia	−78	−31

These differences can be explained by the internal structure of the two types of compound.

Ionic compounds, like sodium chloride, are made of ions. The ions have positive and negative charges and these attract each other very strongly to hold the substance together very tightly. This arrangement of ions is called a **crystal lattice**. A single crystal of sodium chloride contains millions of sodium ions and chloride ions. When crystals of ionic compounds like sodium chloride are heated, a lot of heat energy is needed to overcome the strong attractions (called **ionic bonds**) so that the ions can come apart from each other. When this does happen, the substance melts. In molten sodium chloride, the ions are still attracted to each other, but they are not held rigidly in place.

Covalent compounds like hydrogen chloride gas are made of molecules and not ions. The attraction between the hydrogen and chlorine atoms inside the molecule is strong, but the attraction between the molecules themselves is very weak. This means that hydrogen chloride easily exists as a gas, where the molecules move freely and independently.

Conducting electricity

Ionic compounds conduct electricity when they are molten or when they are dissolved in water. Covalent compounds do not.

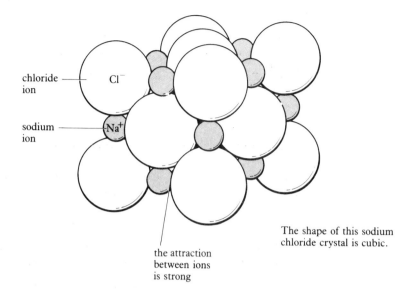

Figure 3
This arrangement of ions is called a crystal lattice. Not all crystals are cubic. What do you think makes crystals of other substances different shapes?

chloride ion

sodium ion

the attraction between ions is strong

The shape of this sodium chloride crystal is cubic.

When ionic compounds are melted, their ions can move around freely in the liquid. Because they have positive and negative charges, electricity is able to flow through the liquid. This process is called **electrolysis** and you can read more about it in section 5.6. Covalent compounds have no ions and so cannot conduct electricity.

Key points

- Elements react in order to get a full outer shell of electrons.
- Some atoms react by gaining or losing electrons and turning into ions. This is called ionic bonding.
- Some atoms share electrons and make uncharged molecules. This is called covalent bonding.
- Information about the charges on ions can be obtained from the periodic table.
- Ionic compounds have higher melting and boiling points than covalent compounds.
- Ionic compounds undergo electrolysis when molten or when dissolved in water. Covalent compounds do not.

Quick questions

1 Write down the electron configurations of the following elements and say which ones will not react.

   (a) Li (b) F (c) Xe (d) K (e) Kr

2 Say whether or not these compounds are likely to be ionic or covalent. Use the position of the elements in the periodic table as a guide.

   (a) lithium chloride, LiCl (b) nitrogen dioxide, NO_2 (c) barium iodide, BaI_2 (d) rubidium oxide, Rb_2O (e) phosphorus(III) bromide, PBr_3.

Questions

1 (a) What is the symbol for phosphorus?
   (b) Name another element in the same group as phosphorus which is a gas.
   (c) Name another element in the same period as phosphorus which is a gas that does not readily form compounds.

SEG

2 An atom has this structure.

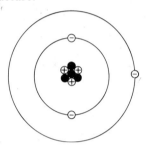

   (a) (i) Name the particle represented by each of the circles:

   (ii) What is represented by ?

   (b) (i) To which group of the Periodic Table does this atom belong?
   (ii) Give the symbol for this element.

SEG

★ 3 Two atoms have the following electronic arrangements.

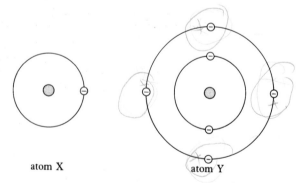

atom X   atom Y

   (a) Draw a diagram to show the electronic arrangement of the compound of X and Y.
   (b) Give the formula of this compound.
   (c) What type of bonding is present in this compound?

SEG

★ **4** Silicon contains three isotopes: $^{28}_{14}Si$, $^{29}_{14}Si$, $^{30}_{14}Si$.

(a) Each atom of silicon contains fourteen electrons. Draw a diagram of the electronic structure of an atom of silicon.

(b) (i) How many protons are present in an atom of $^{30}_{14}Si$?

(ii) How many neutrons are present in an atom of $^{30}_{14}Si$?

(c) What are *isotopes*?

(d) Explain why 28.15 is usually given as the relative atomic mass of silicon.

SEG

★ **5** The table below gives information about four elements, V, W, X and Y, which are in the same group in the Periodic Table.

Element	Atomic number	Melting point/°C	Boiling point/°C
V	9	−220	−188
W	17	−101	−33
X	35	−7	58
Y	53	114	183

(a) Use the melting and boiling points to give the letters of *all* the elements in the table which, at atmospheric pressure and at room temperature, are

(i) solids, (ii) liquids, (iii) gases.

(b) Describe what happens to the particles of a solid when it melts to form a liquid.

(c) The next element in this group after Y is Z.

(i) Would you expect Z to exist under room conditions as a solid, liquid or gas? State your reasons.

(ii) How many electrons will there be in the outer shell of a Z atom?

(iii) Using the symbol Z, write down the formula of the ion of Z.

LEAG

★ **6** An element X forms an ion X^{2-}.

(a) To which group of the Periodic Table does the element X belong?

(b) Write the formula of the compound formed between caesium and X. (Caesium has symbol Cs and is in group 1 of the Periodic Table.)

(c) The ion X^{2-} contains 54 electrons. How many protons are present in the nucleus of an atom of X?

SEG

Using symbols and formulae

4.1 *Formulae*

Insulin

Insulin is a hormone produced by the pancreas, an organ of the body just behind the liver. Insulin is slowly released into the bloodstream where it controls the amount of sugar in the blood. **Diabetes** is a disease in which the amount of sugar in the blood is not properly controlled.

The job performed by insulin was discovered in 1922 when scientists found that extracts from dogs' pancreases relieved the symptoms of diabetes, but it was not until 30 years later that scientists were able to **synthesise** (manufacture) insulin.

Figure 1
Insulin can now be man-made, because its structure is known.

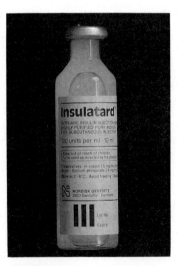

The **empirical formula** of a chemical is the simplest list of elements in the compound and the numbers of each type of atom. The empirical formula of insulin is

$$C_{256}N_{80}H_{515}O_{134}S_{18}$$

This is just a list of symbols and numbers and tells nothing about the way the atoms are joined together in the molecule.

In fact insulin consists of two rows of **amino acids** joined by sulphur atoms. There are 17 different types of amino acid, the simplest of which is **glycine**, which has the formula NH_2CH_2COOH. Before synthetic insulin could be made, scientists had to identify each one of the amino acids and find out the order in which they were joined together.

You can see that a formula needs to tell you much more than just the elements in a compound. It needs to tell you the way in which the atoms are joined. The formulae that follow in this chapter are much simpler than that of insulin. However, even they must be written correctly, otherwise they can give wrong information about the compound.

More ions

Some elements can lose or gain electrons and change into atoms with positive or negative charges. These atoms are called ions. Some examples are:

Na^+	sodium ion	K^+	potassium ion
Ca^{2+}	calcium ion	Al^{3+}	aluminium ion
Cl^-	chloride ion	O^{2-}	oxide ion

Sometimes, groups of atoms can have charges too, and behave as ions. The sulphate ion, SO_4^{2-} is made of one sulphur atom and four oxygen atoms, bonded together. The whole group of atoms has two negative charges.

The nitrate ion, NO_3^-, the carbonate ion, CO_3^{2-}, and the ammonium ion, NH_4^+, are other examples.

Here is a list of most of the ions that you are likely to meet, and will need to know.

Positive ions		Negative ions	
H^+	hydrogen ion	OH^-	hydroxide ion
Li^+	lithium ion	Cl^-	chloride ion
Na^+	sodium ion	Br^-	bromide ion
K^+	potassium ion	I^-	iodide ion
Ag^+	silver ion	NO_3^-	nitrate ion
NH_4^+	ammonium ion		
Ca^{2+}	calcium ion	S^{2-}	sulphide ion
Mg^{2+}	magnesium ion	O^{2-}	oxide ion
Ba^{2+}	barium ion	SO_4^{2-}	sulphate ion
Pb^{2+}	lead ion	SO_3^{2-}	sulphite ion
Cu^{2+}	copper(II) ion	CO_3^{2-}	carbonate ion
Zn^{2+}	zinc ion		
Fe^{2+}	iron(II) ion		
Fe^{3+}	iron(III) ion	PO_4^{3-}	phosphate ion
Al^{3+}	aluminium ion		

Oxidation numbers

Each metal in the table above has an **oxidation number**. It is equal to the number of electrons that the atom lost when it formed an ion. Some metals have more than one ion and then the oxidation number is used to name its compounds. For example iron(II) chloride means the compound $FeCl_2$, which contains the Fe^{2+} ion. But iron(III) chloride means the compound $FeCl_3$, which contains the Fe^{3+} ion.

Writing formulae

In an ionic compound, the ions are attracted to each other by their opposite charges. The number of $+$ charges is always equal to the number of $-$ charges. Knowing this, the formulae of ionic compounds can be written down.

Look at these examples in which the numbers of positive ions and negative ions are balanced to make the charges the same.

Compound	Positive ions	Negative ions	Formula
sodium chloride	Na^+	Cl^-	$NaCl$
calcium chloride	Ca^{2+}	two Cl^- are needed	$CaCl_2$
aluminium chloride	Al^{3+}	three Cl^- are needed	$AlCl_3$
barium sulphate	Ba^{2+}	$SO_4{}^{2-}$	$BaSO_4$
potassium nitrate	K^+	$NO_3{}^-$	KNO_3
lithium carbonate	two Li^+ are needed	$CO_3{}^{2-}$	Li_2CO_3

Now look at this example:

Magnesium hydroxide contains Mg^{2+} and OH^- ions.

The formula is $Mg(OH)_2$. Two OH^- ions are needed to equal the charge on the Mg^{2+} ion. The brackets are needed to show that the 2 refers to the whole OH^- ion and not just the H.

Here are some more examples where brackets are needed.

Compound	Positive ion	Negative ion	Formula
ammonium carbonate	two $NH_4{}^+$	$CO_3{}^{2-}$	$(NH_4)_2CO_3$
copper(II) nitrate	Cu^{2+}	two $NO_3{}^-$	$Cu(NO_3)_2$
aluminium phosphate	two Al^{3+}	three $SO_4{}^{2-}$	$Al_2(SO_4)_3$

Here is a list of rules to follow when you write a formula.

1 Write down the name of the compound.
2 Write down the symbols of the ions involved.

3 Add up the charges of the ions. If they are not equal, add more of the ion that you need. Do this until the charges are equal.
4 Use numbers on the line to show when more than one ion is needed.
5 Use brackets when numbers on the line refer to ions containing more than one atom.

Key points

- If a metal has more than one ion, its oxidation number is used in the names of the metal's compounds.
- When writing a formula, positive and negative ions are put together so that the $+$ and $-$ charges are equal.

Quick question

1 Work out and write down the formula for each of these ionic compounds:

(a) copper(II) sulphate
(b) ammonium carbonate
(c) aluminium nitrate
(d) sodium nitrate
(e) iron(III) sulphate
(f) calcium hydroxide
(g) lithium bromide
(h) iron(II) phosphate
(i) copper(II) bromide
(j) iron(III) phosphate
(k) ammonium phosphate
(l) sodium sulphate

(m) potassium carbonate
(n) zinc carbonate
(o) aluminium oxide
(p) ammonium sulphate
(q) magnesium sulphide
(r) aluminium carbonate
(s) iron(III) chloride
(t) iron(II) sulphite
(u) aluminium hydroxide
(v) iron(II) nitrate
(w) barium oxide
(x) iron(III) sulphite

4.2 Equations

Chemical sentences?

| t e h a c s |
| n o m |

letters

| H O C |
| Cl Cu |

symbols

| mat cat the |
| on sat |

words

| $CuCO_3$ HCl CO_2 |
| $CuCl_2$ H_2O |

formulae

| the cat sat on the mat |

sentence

| $CuCO_3 + 2HCl \longrightarrow CuCl_2 + CO_2 + H_2O$ |

equation

Are spelling and grammar important when you write an English essay? You could ask the same question about writing chemical formulae and equations.

Equations

Equations are chemical sentences. They explain what is happening in chemical reactions. They also tell us how much of the different chemicals are reacting with each other.

In order to write balanced equations, you should follow this set of rules:

1 Write down the chemical reaction in words. Make sure that all the chemicals are there.

$$reactants \longrightarrow products$$

2 Write down the symbols or formulae of each of the elements or compounds involved in the reaction. Make sure they are absolutely correct.

3 Balance the equation so that there are exactly the same numbers of atoms of each element on both sides.

Example 1

$$iron + sulphur \longrightarrow iron(II) \ sulphide$$

This is the reaction that takes place when iron filings are heated with sulphur powder. Black iron(II) sulphide is formed. Here are the symbols and formula:

$$Fe + S \longrightarrow FeS$$

Now count the numbers of each atom

$$\begin{matrix} 1 \ atom \\ of \ Fe \end{matrix} + \begin{matrix} 1 \ atom \\ of \ S \end{matrix} \longrightarrow 1 \ atom \ of \ Fe, \ 1 \ atom \ of \ S$$

The numbers of atoms of each elements are the same on both sides of the equation, so the equation is balanced.

The symbols of state can now be added. They show if the substance is a solid (s), liquid (l), a gas (g), or dissolved in water (aq).

$$Fe(s) + S(s) \longrightarrow FeS(s)$$

Example 2

When magnesium is put into dilute hydrochloric acid, it fizzes and hydrogen gas is released. A solution of magnesium chloride is left.

$$magnesium + \begin{matrix} hydrochloric \\ acid \end{matrix} \longrightarrow \begin{matrix} magnesium \\ chloride \end{matrix} + hydrogen$$

Now for the symbols and formulae:

$$Mg + HCl \longrightarrow MgCl_2 + H_2$$

Make sure that the symbols for hydrochloric acid and magnesium chloride are correct. Remember that gases like hydrogen exist as molecules of two atoms.

Now count the atoms:

$$\text{1 atom of Mg} + \begin{matrix}\text{1 atom of H}\\\text{1 atom of Cl}\end{matrix} \longrightarrow \begin{matrix}\text{1 atom of Mg}\\\text{2 atoms of Cl}\end{matrix} + \text{2 atoms of H}$$

The numbers of atoms are not the same on each side of the equation. The symbols and formulae must not be changed, but you can have more than one molecule of each substance in the equation. The problem is solved by adding an extra molecule of hydrochloric acid.

$$Mg + 2HCl \longrightarrow MgCl_2 + H_2$$

The equation is now balanced.

Finally, put in the symbols of state.

$$Mg(s) + 2HCl(aq) \longrightarrow MgCl_2(aq) + H_2(g)$$

Example 3

When sodium hydroxide solution is added to dilute sulphuric acid, a neutralisation reaction takes place. A salt called sodium sulphate is made.

$$\begin{matrix}\text{sodium}\\\text{hydroxide}\end{matrix} + \begin{matrix}\text{sulphuric}\\\text{acid}\end{matrix} \longrightarrow \begin{matrix}\text{sodium}\\\text{sulphate}\end{matrix} + \text{water}$$

$$NaOH + H_2SO_4 \longrightarrow Na_2SO_4 + H_2O$$

Now count the atoms on each side. There are not enough sodium atoms on the left hand side. Solve the problem by adding another molecule of sodium hydroxide.

$$2NaOH + H_2SO_4 \longrightarrow Na_2SO_4 + H_2O$$

Now there are not enough hydrogen and oxygen atoms on the right hand side. Solve this problem by adding another molecule of water. Finally, add the symbols of state.

$$2NaOH(aq) + H_2SO_4(aq) \longrightarrow Na_2SO_4(aq) + 2H_2O(l)$$

Example 4

See if you can follow this sequence of events.

When potassium carbonate is added to dilute nitric acid, potassium nitrate solution is formed and carbon dioxide gas is released.

$$\begin{matrix}\text{potassium}\\\text{carbonate}\end{matrix} + \begin{matrix}\text{nitric}\\\text{acid}\end{matrix} \longrightarrow \begin{matrix}\text{potassium}\\\text{nitrate}\end{matrix} + \begin{matrix}\text{carbon}\\\text{dioxide}\end{matrix} + \text{water}$$

$$K_2CO_3 + HNO_3 \longrightarrow KNO_3 + CO_2 + H_2O$$
$$K_2CO_3 + 2HNO_3 \longrightarrow 2KNO_3 + CO_2 + H_2O$$
$$K_2CO_3(aq) + 2HNO_3(aq) \longrightarrow 2KNO_3(aq) + CO_2(g) + H_2O(l)$$

Key points

- Equations are balanced by having the same number of each type of atom on each side.
- Symbols of state tell you about the nature of a substance.

Quick question

* 1 Write equations for these reactions. Put in the symbols of state if you can.

(a) zinc + sulphur ⟶ zinc sulphide

(b) copper + chlorine ⟶ copper(II) chloride
(remember that the formula for chlorine is Cl_2)

(c) sodium + chlorine ⟶ sodium chloride

(d) calcium + iodine ⟶ calcium iodide

(e) potassium hydroxide + nitric acid ⟶ potassium nitrate + water

(f) lithium hydroxide + sulphuric acid ⟶ lithium sulphate + water

(g) zinc + hydrochloric acid ⟶ zinc chloride + hydrogen

(h) iron + hydrochloric acid ⟶ iron(II) chloride + hydrogen

(i) calcium carbonate + nitric acid ⟶ calcium carbonate + carbon dioxide + water

(j) copper + oxygen ⟶ copper(II) oxide

(k) hydrogen + oxygen ⟶ steam

(l) methane + oxygen ⟶ carbon dioxide + water

(m) aluminium oxide + hydrochloric acid ⟶ aluminium chloride + water

4.3 *Calculations*

Relative molecular mass

Each element has its own **relative atomic mass**. This is the average mass of its isotopes compared with the mass of a standard atom of carbon, $^{12}_{6}C$. There is a list of relative atomic masses at the back of the book.

Compounds have a **relative molecular mass** (sometimes called **formula mass**). The relative molecular mass (formula mass) of a compound is found by adding up the relative atomic masses of the elements in the compound according to the number of each element's atoms.

Look at these examples:

1 Sodium chloride, NaCl

1 atom of Na	1×23	=	23
1 atom of Cl	1×35.5	=	35.5 +
relative molecular mass		=	58.5

2 Calcium chloride, $CaCl_2$

1 atom of Ca	1×40	$= \quad 40$
2 atoms of Cl	2×35.5	$= \quad 71 +$
relative molecular mass		$= \quad 111$

3 Zinc nitrate, $Zn(NO_3)_2$

1 atom of Zn	1×65	$= \quad 65$
2 atoms of N	2×14	$= \quad 28$
6 atoms of O	$2 \times 3 \times 16$	$= \quad 96 +$
relative molecular mass		$= \quad 189$

Percentage composition by mass

Sometimes it is important to know the exact composition of a compound. For example, all bags of fertilizer must show the percentage of the elements nitrogen, phosphorus and potassium they contain on the outside.

$$\text{The \% mass of an element in a compound} = \frac{\text{relative atomic mass of the element in the compound}}{\text{the relative molecular mass of the compound}}$$

This calculation will show the proportion of any type of atom in a compound.

Look at these examples:

1 The % mass of calcium in calcium carbonate, $CaCO_3$

$$= \frac{Ca}{Ca + C + 3 \times O} \times 100 = \frac{40}{40 + 12 + 48} \times 100 = 40\%$$

40% of the mass of calcium carbonate is calcium.

2 The % mass of nitrogen in ammonium nitrate, NH_4NO_3

$$= \frac{2 \times N}{2 \times N + 4 \times H + 3 \times O} \times 100 = \frac{28}{28 + 4 + 48} \times 100 = 35\%$$

28% of the mass of ammonium nitrate is nitrogen.

3 The % mass of water in copper(II) sulphate crystals, $CuSO_4.5H_2O$

$$= \frac{5 \times H_2O}{Cu + S + 4 \times O + 5 \times H_2O} \times 100 = \frac{5 \times 18}{64 + 32 + 64 + 90} \times 100$$
$$= 36\%$$

36% of the mass of copper(II) sulphate crystals is water.

The mole

Chemists measure the amount of substances in **moles**. The word 'mole' comes from a Latin word meaning a heap or a pile, so a mole is a heap or pile of atoms or molecules.

> A mole of any substance is the amount of that substance that contains as many atoms, or molecules, as there are atoms in exactly 12 g of the isotope $^{12}_{6}C$.

A mole has no units. It is not measured in grams. It is just an amount of matter.

The actual number of atoms or molecules in a mole is very large. There are 602 000 000 000 000 000 000 000 or 6.02×10^{23} of them. This large number has the symbol L and is called the **Avogadro constant**.

Molar mass

Chemists measure masses of chemicals in **molar masses**.

> A molar mass of a substance is the mass of a mole of the substance and is same as a mole, measured in grams.

Many chemists use the word mole when they mean molar mass. They say:

> A mole (molar mass) of any substance is its relative atomic mass, or relative molecular mass (if it is a compound) measured in grams.

Look at these examples:

1 1 mole of sulphur = the relative atomic mass (r.a.m.) in grams
$$= 32 \text{ g}$$

2 2 moles of calcium = 2×40 g
$$= 80 \text{ g}$$

3 1 mole of water = the relative molecular mass of water in grams
$$= 2 \times H + O$$
$$= 2 + 16$$
$$= 18 \text{ g}$$

These calculations can be done the other way round too:

4 How many moles of calcium carbonate are there in 200 g of the substance?

$$1 \text{ mole of } CaCO_3 = Ca + C + 3 \times O$$
$$= 40 + 12 + 48$$
$$= 100 \text{ g}$$

So, in 200 g of $CaCO_3$, there are $\dfrac{200}{100}$ moles = 2

There are 2 moles in 200 g of calcium carbonate.

Calculations from equation

Not only do equations tell you what substances are reacting and being formed in a chemical reaction, but they tell you the amounts of the substances reacting.

The equation shows how many moles of each substance are reacting.

Example 1

$$\text{iron} + \text{sulphur} \longrightarrow \text{iron(II) sulphide}$$
$$\text{Fe(s)} + \text{S(s)} \longrightarrow \text{FeS(s)}$$

This means

$$\text{1 mole} + \text{1 mole} \longrightarrow \text{1 mole}$$

and in grams

$$56\,\text{g} + 32\,\text{g} \longrightarrow 88\,\text{g}$$

Example 2

$$\text{magnesium} + \text{hydrochloric acid} \longrightarrow \text{magnesium chloride} + \text{hydrogen}$$
$$\text{Mg(s)} + \text{2HCl(aq)} \longrightarrow \text{MgCl}_2\text{(aq)} + \text{H}_2\text{(g)}$$

this means

$$\text{1 mole} + \text{2 moles} \longrightarrow \text{1 mole} + \text{1 mole}$$

and in grams

$$24\,\text{g} + 73\,\text{g} \longrightarrow 95\,\text{g} + 2\,\text{g}$$

In each of the examples so far, the total mass on the left-hand side of the equation is equal to the total mass on the right-hand side of the equation. Go back and check it!

Equations can be used to predict the amount of a substance that will be used up or made in a chemical reaction. Look at the examples:

Example 3

How much calcium chloride can be made when 80 g of calcium is dissolved in dilute hydrochloric acid?

Follow this routine:

1 Write the equation:

$$\text{calcium} + \text{hydrochloric acid} \longrightarrow \text{calcium chloride} + \text{hydrogen}$$
$$\text{Ca(s)} + \text{2HCl(aq)} \longrightarrow \text{CaCl}_2\text{(aq)} + \text{H}_2\text{(g)}$$

2 Now pick out the substances involved in the question:

$$\text{Ca(s)} \longrightarrow \text{CaCl}_2\text{(aq)}$$

3 How many moles are there of each?

$$\text{1 mole} \longrightarrow \text{1 mole}$$

4 Now change moles into grams

$$40 \text{ g} \longrightarrow 111 \text{ g}$$

5 So, if 40 g of Ca will make 111 g of $CaCl_2$
then, 80 g of Ca will make 222 g of $CaCl_2$.

222 g of calcium chloride will be made.

Example 4

How much hydrochloric acid is needed to make 100 g of copper(II) chloride from copper(II) oxide?

Follow the same routine:

1 copper(II) oxide $+$ hydrochloric acid $\longrightarrow$ copper(II) chloride $+$ water

$$CuO(s) \ + \ 2HCl(aq) \ \longrightarrow CuCl_2(aq) + H_2O(l)$$

2 $\qquad\qquad 2HCl(aq) \longrightarrow CuCl_2(aq)$

3 $\qquad\qquad$ 2 moles $\longrightarrow$ 1 mole

4 $\qquad\qquad\qquad$ 73 g $\longrightarrow$ 135 g

So 135 g of $CuCl_2$ can be made from 73 g of acid,

and 1 g of $CuCl_2$ can be made from $\dfrac{73}{135}$ g of acid,

and 100 g of $CuCl_2$ can be made from $\dfrac{100 \times 73}{135}$ g of acid

$\dfrac{73 \times 100}{135}$ g of copper(II) chloride can be made $= 59.5$ g

Example 5

How much ammonia can be made from 1 kilogram of hydrogen?

$$\text{hydrogen} + \text{nitrogen} \longrightarrow \text{ammonia}$$
$$3H_2(g) \ + \ N_2(g) \ \longrightarrow \ 2NH_3(g)$$
$$3H_2(g) \longrightarrow 2NH_3(g)$$
$$3 \text{ moles} \longrightarrow 2 \text{ moles}$$
$$6 \text{ g} \longrightarrow 34 \text{ g}$$
$$6 \text{ kg} \longrightarrow 34 \text{ kg}$$
$$1 \text{ kg} \longrightarrow 5.7 \text{ kg}$$

5.7 kg of ammonia will be made.

Using experiments to find formulae

Reducing lead(II) oxide

1 Weigh a sample of black lead(IV) oxide and put it into a small porcelain dish in a reduction tube. (See Figure 1.)
2 Turn on the hydrogen for several seconds before lighting it at the jet at the end of the tube.
3 Gently heat the lead(IV) oxide while the hydrogen flows over it until the black powder changes to silver beads of lead.
4 When the apparatus has cooled, turn off the hydrogen and weigh the lead.

Figure 1
The hydrogen is allowed to flow for several seconds before it is lit. Why is this an important safety precaution? At the end of the experiment, the lead is allowed to cool before the flow of hydrogen is stopped. What would happen if air got to the hot lead?

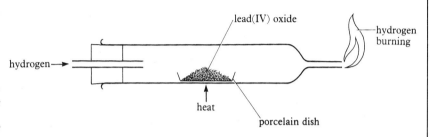

Here are some sample results:

mass of dish   = 26.00 g
mass of dish and lead(IV) oxide   = 28.39 g
mass of dish and lead   = 28.07 g

mass of lead on its own

$$
\begin{array}{r}
28.07 \text{ g} \\
- 26.00 \text{ g} \\
\hline
= 2.07 \text{ g}
\end{array}
$$

mass of oxygen in the oxide

$$
\begin{array}{r}
28.39 \text{ g} \\
- 28.07 \text{ g} \\
\hline
= 0.32 \text{ g}
\end{array}
$$

So, 2.07 g of lead was combined with 0.32 g of oxygen in the lead(IV) oxide.

These masses can be changed into moles by dividing by the molar mass.

$$2.07 \text{ g of lead} = \frac{2.07}{207} = 0.01 \text{ mole}$$

$$0.16 \text{ g of oxygen} = \frac{0.32}{16} = 0.02 \text{ mole}$$

In other words, lead(IV) oxide contains twice the number of moles of oxygen as lead, so the simplest, or empirical formula of lead(IV) oxide is PbO_2.

Imagining a mole

The Avagadro constant is an enormous number.

$$L = 6.02 \times 10^{23}$$

Here are some ideas to help you imagine just how big this number is.

1 If you had this number of grains of sand, you could spread them out over the surface of Great Britain and they would be over 20 metres deep.
2 If you had this number of sheets of paper and you made them into a million equal piles, each pile would reach beyond the sun.
3 If you had a computer that could count 10 million times a second, it would take 2 billion years to count Avogadro's number.

Key points

- Each compound has a relative molecular mass, sometimes called formula mass.
- The percentage by mass of any element in a compound may be calculated.
- A mole of any substance is an exact number of atoms or molecules.
- A mole of any substance, measured in grams, is called the molar mass of that substance.
- The formula of a substance can be obtained by experiment.

Quick question

1 Calculate the relative molecular masses (formula masses) of the following compounds. You will have to write down the correct formula first of all.

(a) potassium carbonate (d) aluminium bromide
(b) iron(III) hydroxide (e) sulphuric acid
(c) zinc nitrate   (f) calcium hydroxide.

★ 2 Calculate the percentage mass of

(a) sulphur in sulphur dioxide (b) hydrogen in methane (c) oxygen in copper(II) nitrate (d) chlorine in calcium chloride (e) sodium in sodium sulphate (f) water in $Na_2CO_3.10H_2O$.

★ 3 Calculate

(a) 1 mole of nitric acid (b) 2 moles of sodium hydroxide
(c) 0.5 moles of carbon dioxide (d) 10 moles of water.

★ 4 How many moles are there in

(a) 60 g of carbon (b) 100 g of neon (c) 5.6 g of iron (d) 72 g of water.

★ 5 Use this equation

$$CuO(s) + H_2(g) \longrightarrow Cu(s) + H_2O(g)$$

to calculate how much copper could be made from 320 g of copper(II) oxide.

* **6** Use this equation

$$CH_4(g) + 2O_2(g) + 2H_2O(g)$$

to calculate how much methane has to be burned to make 180 g of steam.

* **7** When magnesium is burned in oxygen, 12 g of magnesium is found to have formed 20 g of magnesium oxide. Calculate the empirical formula of the oxide.

Questions

* **1** Excess dilute nitric acid was reacted with 2 g of magnesium oxide. The equation for this reaction is

$$MgO + 2HNO_3 \longrightarrow Mg(NO_3)_2 + H_2O$$

   (i) Describe a simple experiment to show that the nitric acid was in excess.
   (ii) How many moles of nitric acid, HNO_3, react with 1 mole of magnesium oxide, MgO?
   (iii) The mass of 1 mole of MgO is 40 g. Calculate the number of moles of MgO present in the 2 g of MgO used in this reaction.
   (iv) How many moles of HNO_3 react with 2 g of MgO?
   (v) What mass of HNO_3 reacts with 2 g of MgO?
   (The mass of 1 mole of HNO_3 is 63 g.)
   (vi) If the dilute acid used contains 1 mole of HNO_3 in 1000 cm^3 of solution, what volume of the acid reacts with 2 g of MgO?

NEA

* **2** A sample of corrosion scraped from the inside of a water heater had a mass of 5.8 g. When it was analysed, it was found to contain 4.2 g of iron, and the rest was oxygen. Calculate the simplest formula of the compound.

* **3** The ore from which aluminium is extracted is called Bauxite. Its formula is $Al_2O_3.3H_2O$.
   (a) What is the formula mass of Bauxite?
   (b) What is the percentage by mass of aluminium in Bauxite?
   (c) Calculate how much aluminium could be extracted from 10 kg of Bauxite.

* **4** Two compounds are used as fertilizers. They are ammonium nitrate $[NH_4NO_3]$, and ammonium sulphate $[(NH_4)_2SO_4]$.
   Say which one is the best fertilizer by calculating the percentage of nitrogen in each compound.

5 Reactions

5.1 *Chemical reactions*

Fireworks

Bangers bang and Catherine wheels spin, sparks fly and bright lights flare, as a result of **chemical reactions**. The basic ingredient of most fireworks is gunpowder. The Chinese had it more than 800 years ago, long before guns were invented. Gunpowder consists of potassium nitrate, carbon and sulphur, all ground up into a fine powder. When it is set alight, the potassium nitrate decomposes to form oxygen. The oxygen enables the carbon and sulphur to burn and produce carbon dioxide and sulphur dioxide gases. If the mixture is very finely powdered then the reaction will happen quickly. If it is put into a cardboard tube, the expanding gases will rip the tube apart and produce a shock wave that we hear as a bang. In a cartridge, this energy is used to propel a bullet.

To make the reaction go even faster, other oxygen containing compounds are added along with sawdust and resins. Powdered magnesium, aluminium or titanium and compounds of strontium, barium, sodium and potassium colour the flames steely white, red, green, yellow. Rockets are made by packing the mixture into a tube in such a way that the gases produced push down the tube they are in. The inside of the rocket has to be hollow for this to happen and a stick must be added to keep the rocket going in a straight line.

Making fireworks is a dangerous business and it is against the law to make your own. People who do make fireworks professionally take precautions. Machines are avoided if possible because of the danger of heat and sparks. Even shops selling fireworks have to follow rules. They may only sell them three weeks before November 5th and no more than 50 kg may be kept in the shop at any one time. A maximum of 1000 kg may be stored outside.

Figure 1
*What chemical reactions
are shown taking place
in those photographs? Are
the reactions fast or slow?*

Chemical reactions

When chemical reactions take place, new substances are formed. The substances reacting are called **reactants** and the new substances being formed are called **products**. The atoms and molecules in the reactants rearrange themselves to make new substances. In all chemical reactions, energy is involved. Sometimes it is needed to start the reaction and often it is given out during the reaction.

The products of a reaction are quite different to the reactants and it is often difficult or impossible to change them back again. Look at these three examples of chemical reactions.

Example 1

When a Bunsen burner is lit, a chemical reaction takes place. Heat energy in the form of a flame, or spark, is needed to start the reaction. Then, in the mixture of air and gas, molecules of oxygen and methane split apart to form new molecules of carbon dioxide and steam. These are the products of the reaction. The flame is, in fact, a hot reaction.

$$CH_4(g) + 2O_2(g) \longrightarrow CO_2(g) + 2H_2O(g)$$

Example 2

When a piece of zinc is put into dilute hydrochloric acid, the zinc reacts forming zinc chloride solution. Hydrogen is released. The chemicals get hot too.

$$Zn(s) + 2HCl(aq) \longrightarrow ZnCl_2(aq) + H_2(g)$$

Example 3

When a piece of magnesium ribbon is heated in air, atoms of magnesium and molecules of oxygen change into white magnesium oxide which is a crystal lattice of ions.

$$2Mg(s) + O_2(g) \longrightarrow 2MgO(s)$$

Different forms of energy

The energy given out in some chemical reactions is sometimes important too. The light energy coming from a torch bulb comes from the electricity made when the chemicals in the battery react. The pellets in a shotgun cartridge are forced out by the hot expanding gases formed when the gunpowder burns. The metal in the photograph is being cut by the intense heat produced when a mixture of ethyne (acetylene) and oxygen burns.

Not all changes in appearance are a sign that a chemical reaction has taken place though! The ice in the photograph is melting but this isn't a chemical reaction. Nothing new is being made. Ice and water are the same chemical substance. Their molecules are the same. They are just arranged differently. Changes like melting, boiling, condensing and freezing are

physical changes. Dissolving, distilling and crystallising are physical changes too. Mixtures are made or separated, there is no permanent change.

Figure 2 *Why is this change not a chemical reaction? How would you change the water back to ice again?*

Figure 3 *Pipes lead gas into the end of this burner. Why are there two of them?*

Key points

- When chemical reactions take place, new substances are formed which cannot easily be changed back to the original substances.
- Energy given out during chemical reactions can be useful.
- Changes like dissolving, melting and boiling are not chemical reactions, but physical changes.

Quick question

1 Sort the following into chemical reactions and physical changes. Give your reasons.
   (a) iron rusting
   (b) a window misting up in a warm room
   (c) a match burning
   (d) a penny going brown
   (e) sugar dissolving in tea

5.2 *Changing the speed of reactions*

Exploding flour

On 18th October 1982, the inhabitants of Metz in France were shocked by a loud explosion. People near the local flour mill saw flames coming from the top of a barley silo 70 metres tall and had to duck as large chunks of concrete were flung by the blast onto a motorway 250 metres away. The cause was dust!

Figure 1
This flour mill was wrecked by a dust explosion.

If tiny particles of flour or grain dust get very hot, they burn. The smaller they are, the more easily they can be suspended in the air as a fog. If the density of this fog is between 0.02 and 2 kg of dust per cubic metre of air, burning spreads so rapidly that a chain reaction can build up in the particles. The temperature rises quickly and a huge volume of gas (carbon dioxide and steam) is produced in a very short space of time. This causes a shock wave that can blow apart the container that the dust is in. All that is needed to start it off is a spark or a hot surface. This means that flour mills can be very dangerous places unless special precautions like these are taken.

Dust must be kept to a minimum. Doors between rooms must be dust tight and valves and openings on machines too. Some rooms must be kept at a pressure lower than the outside so that dust cannot get out. No naked flames can be allowed. Dust must not come into contact with hot surfaces. Sparks from friction or static electricity must be prevented and stored grain must not be allowed to heat up.

Smaller particles — faster reaction

- Dust in flour mills can explode.
- Tablets dissolve more easily if they are ground up into a powder.
- Sawdust burns more easily than tree trunks.

Do small particles react more quickly than big lumps?

The flask contains dilute hydrochloric acid and marble chips. Each chip is about 1 cm across.

The reaction

$$CaCO_3(s) + 2HCl(aq) \longrightarrow CaCl_2(aq) + CO_2(g) + H_2O(l)$$

takes about 45 seconds to produce enough carbon dioxide to fill the syringe. If powdered marble is used instead, the syringe fills in about 10 seconds.

Powdered marble chips react faster than lumps.

Figure 2
What other piece of apparatus will you need to measure how fast carbon dioxide is being given off?

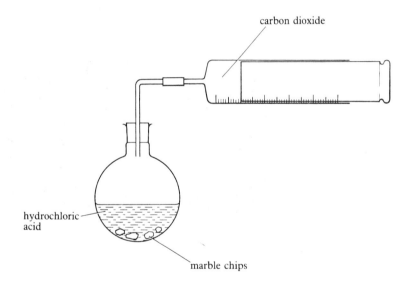

carbon dioxide

hydrochloric acid

marble chips

When a reaction takes place between a solid and a liquid (or a gas), the amount of surface area is important. In the experiment with powdered marble chips, the powder has more surface area for the acid to react with, so the reaction is quicker.

Higher temperature — more energy

Many chemical reactions need heat energy to start them, and the hotter they get, the faster the reaction goes. Look at this experiment.

The reaction of zinc with sulphuric acid

$$Zn(s) + H_2SO_4(aq) \longrightarrow ZnSO_4(aq) + H_2(g)$$

Bubbles of hydrogen come off the zinc at a rate of about one per second. At a higher temperature, the bubbles come off much faster.

Higher temperatures mean faster reactions.

Figure 3
What gas is being produced in this reaction? What must be done to the chemicals in the flask to make the gas bubble faster?

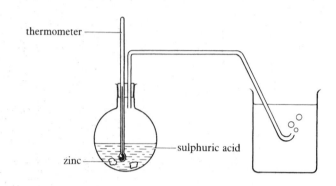

thermometer

sulphuric acid

zinc

Higher concentration — faster reaction

Acids are diluted by mixing with water. The less water there is, and the more acid, the greater the concentration of the acid becomes. Dilute acids react slowly with metals. If the acid is made more concentrated, the reaction takes place faster. Look at this experiment.

The reaction of magnesium with sulphuric acid

$$Mg(s) + H_2SO_4(aq) \longrightarrow MgSO_4(aq) + H_2(g)$$

Using ordinary laboratory dilute acid, it takes 20 seconds to fill the test tube with hydrogen. If the acid is diluted even more by adding water, the reaction is slower.

Figure 4
What substance could be added to the acid in the flask to slow the reaction down?

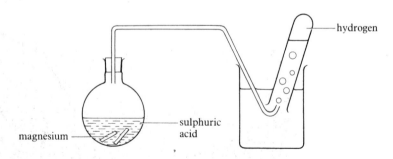

hydrogen

magnesium

sulphuric acid

For gases, increasing the pressure has the same effect as increasing the concentration. Because the molecules are squashed closer together there are more molecules in the same space. A higher pressure means a faster reaction.

Oxygen under high presure in a cylinder can react explosively with oil.

Figure 5
*If oil **were** put into this valve, it would mix with oxygen at high pressure. Why would this be dangerous?*

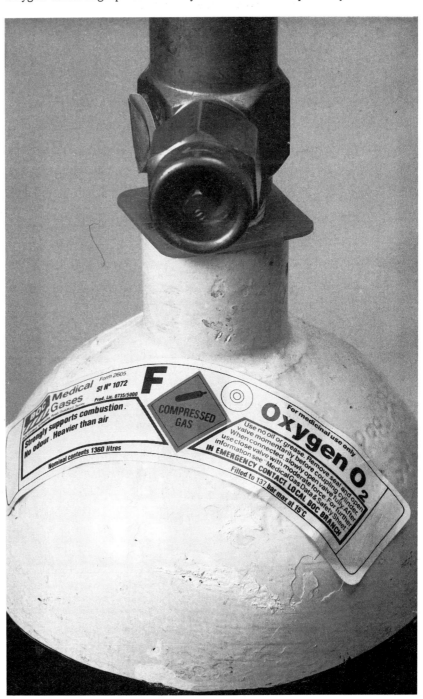

The Collision Theory

You stand a greater chance of being bumped into

    1 if there are lots of people around
    2 if they are all running instead of walking.

The same is true of particles (atoms, molecules or ions). Chemical reactions take place when particles of reactants collide. The collision provides the energy to break bonds and make new ones. The speed of the reaction depends upon how often the particles collide, and this idea is called **the collision theory**.

The speed of the reaction depends upon:

1 how many particles there are (in other words, how concentrated the substances are).
2 how fast they are moving (in other words, how hot they are).
3 how close the particles are to each other if the substance is a gas (in other words, how high the pressure is).
4 how big the surface area is of the substance reacting (the larger the surface area, the more particles become exposed).

Light

Some chemical reactions get the energy they need to react from light. They are called **photochemical reactions**.

Silver nitrate solution is often kept in brown glass bottles because light makes it decompose into silver. Photographic films are made from silver bromide. This chemical is changed into silver by light.

Figure 6   *The dark parts on this film are where silver has been formed. The clear parts are where unreacted silver bromide has been removed in the processing. Why do you think this is called a negative?*

Photosynthesis is a biological reaction in plants in which water and carbon dioxide are changed into oxygen and sugar. A catalyst called chlorophyll is needed and the reaction needs light to work. You can read more about photosynthesis in section 7.2.

Catalysts

Exhaust from car engines contains harmful gases like carbon monoxide, oxides of nitrogen, and unburnt hydrocarbons, as well as harmless carbon dioxide and steam. In the USA and Japan, the amount of these poisonous gases allowed to come out of exhaust pipes is controlled by law. All cars in these countries have to have **catalyst converters** fitted in their exhaust systems. These consist of a ceramic honeycomb, with thousands of tiny holes in it, through which the exhaust gases pass. The tiny holes are coated with a thin layer of catalyst — a mixture of platinum and rhodium.

This catalyst mixture makes chemical reactions take place:

carbon monoxide	turns into carbon dioxide
oxides of nitrogen	turn into nitrogen
unburnt hydrocarbons	turn into steam and carbon dioxide.

A catalyst is a substance which:
1 makes a reaction that would only happen slowly on its own, go faster.
2 does not get used up itself.

You can read more about catalysts used in industry in sections 7.1–7.9.

Using a catalyst to make oxygen in the laboratory

At room temperature, hydrogen peroxide decomposes very slowly.

$$2H_2O_2(aq) \longrightarrow 2H_2O(l) + O_2(g)$$

If a catalyst called manganese(IV) oxide is added, decomposition takes place very vigorously. After the hydrogen peroxide has finished fizzing, the manganese(IV) oxide is not used up.

Figure 7
The catalyst in this reaction had a mass of 20 g when it was put in. What will be its mass after the reaction has finished?

Key points

- Dust particles can cause explosions.
- Reactions can be speeded up by increasing concentration, pressure, temperature; by using more finely divided particles, and in some cases, with light.
- The way that reactions are speeded up can be explained by the Collision Theory.
- Catalysts are substances which speed up some chemical reactions, but do not get used up themselves.

Quick questions

1 Some indigestion tablets fizz and dissolve in water. Describe three ways in which you might make them dissolve more quickly.

2 If a chemical is said to decompose photochemically, what does this mean? How should the chemical be stored?

3 Describe how you might show whether or not iron powder is a catalyst for the decomposition of hydrogen peroxide solution.

★ 4 A black powder, P, is put into a solution, Q. A fizzy reaction takes place after which the mixture is filtered. The residue is dried and weighed. The filtrate is weighed too. These are the results that were obtained.

mass of flask	= 80.00 g
mass of flask + P	= 85.00 g
mass of Q	= 10.00 g
mass of dried residue	= 5.00 g
mass of filtrate	= 7.00 g

(a) Do you think P is a catalyst? Use the weighings to support your answer.

(b) Why do you think the filtrate weighs less than Q did at the start?

5.3 *Energy in reactions*

Fuels for space

All rockets work in the same way. A fuel is burned very quickly so that a huge volume of hot gas is produced in a small space in a short time. This chemical reaction takes place inside a **combustion chamber**. The chamber has an opening at one end so the gases can escape quickly, pushing the rocket in the opposite direction.

This is the reaction that takes place in the combustion chamber.

fuel + oxygen ———→ hot expanding gases

For the reaction to take place, oxygen must be present. The fuels in rockets burn very quickly and use so much oxygen that the oxygen in the air alone would not be enough for the combustion. So, as well as fuel, the rocket must carry its own oxygen supply, or a compound that contains oxygen. A substance that contains oxygen that will help a fuel to burn is called an **oxidant.**

Figure 1
A rocket combustion chamber. The hot gases escape through the nozzle and the rocket is pushed in the opposite direction. Why will the shape of the nozzle be important?

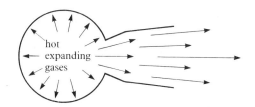

Figure 2
Most of this rocket is full of fuel. Why must it also contain oxygen or oxidants? (there are 2 reasons!)

The giant Saturn rockets which took the Apollo missions, and first took men to the moon in 1969, were made of three separate parts or stages.

Stage 1 was needed to get the massive bulk of the rocket off the ground and 58 km into space, at a speed of 9600 km per hour. It used kerosine (paraffin) and liquid oxygen as an oxidant and burned 15 tonnes of these liquids every second!

Once the fuel in stage 1 had burned away, the stage fell away, leaving **Stage 2**, powered by liquid hydrogen and liquid oxygen to get the remainder of the rocket into Earth's orbit.

Finally, **Stage 3**, also using liquid hydrogen and oxygen, took the men on to the moon.

Exothermic and endothermic reactions

All chemical reactions involve an energy change. Often, this energy is in the form of heat.

When sodium hydroxide solution is added to dilute hydrochloric acid, a **neutralisation** reaction takes place. A solution of sodium chloride is formed.

$$NaOH(aq) + HCl(aq) \longrightarrow NaCl(aq) + H_2O(l)$$

The liquids get hot when they are mixed. This is called an **exothermic** reaction because heat energy is given out.

Here are some more examples of exothermic reactions:

burning methane (North Sea Gas),
burning magnesium ribbon,
diluting concentrated sulphuric acid.

When blue copper sulphate crystals are heated, they change into a white powder and steam is given off.

$$CuSO_4.5H_2O(s) \longrightarrow CuSO_4(s) + 5H_2O(g)$$

Heat has to be taken in to make the reaction work. This is called an **endothermic** reaction.

Here are some more endothermic reactions:

adding ammonium carbonate to ethanoic acid,
dissolving copper sulphate crystals in water.

> Exothermic reactions give out heat energy.
> Endothermic reactions take in heat energy.

Measuring heat energy

Heat energy given out, or taken in, during a reaction is measured in units called joules (J). Large amounts of heat energy are measured in kilojoules.

$$1000 \text{ joules} = 1 \text{ kilojoule} \qquad 1000 \text{ J} = 1 \text{ kJ}$$

The symbol for the heat energy change when a reaction takes place is ΔH.

> If heat is given out (exothermic), heat is lost. ΔH is negative.
>
> If heat is taken in (endothermic), heat is gained. ΔH is positive.

Here are two examples.

Example 1; exothermic

$$NaOH(aq) + HCl(aq) \longrightarrow NaCl(aq) + H_2O(l)$$

The heat energy change for this reaction is -57.5 kJ per mole of sodium hydroxide. Heat is generated and the liquids get hot.

$$\Delta H = -57.5 \text{ kJ mol}^{-1}$$

Example 2; endothermic

$$H_2 \longrightarrow 2H$$

The heat energy change when molecules of hydrogen are split into atoms is $+436$ kJ per mole of hydrogen molecules.

$$\Delta H = +436 \text{ kJ mol}^{-1}$$

Hydrogen gas has to be made very hot indeed to provide enough energy for this to happen.

Energy diagrams

Heat energy changes in reactions can be represented by diagrams. Figure 3 shows an exothermic reaction. The sodium chloride and water have a lower energy level than the sodium hydroxide and hydrochloric acid, so heat energy is given out when the reaction takes place.

Figure 4 shows hydrogen atoms have a higher energy level than hydrogen molecules. Heat energy is taken in when this endothermic reaction takes place.

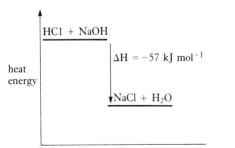

Figure 3 *Heat is given out when an acid is neutralized by an alkali. This makes the reactants hot.*

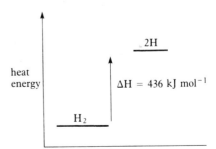

Figure 4 *Heat energy must be taken in to make hydrogen molecules split into atoms.*

Comparing the heat energy of fuels

When a fuel burns, heat energy is given out. The amount of heat energy depends upon the sort of bonds in the fuel that have to be broken when the fuel reacts with oxygen. Different fuels have different heat energy values. They give out different amounts of heat.

---

*Experiment to show how the **heat energy value** (sometimes called **heat of combustion**) of a fuel could be measured.*

Figure 5
Why are the insulation and screens necessary? How will the heat of combustion of each fuel affect the temperature of the water at the end of the experiment?

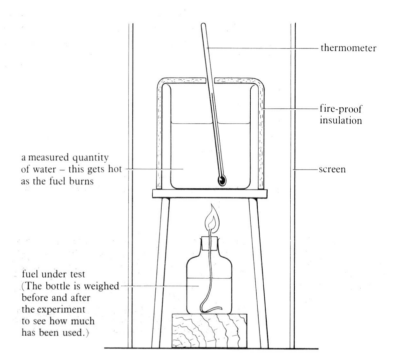

thermometer

fire-proof insulation

a measured quantity of water – this gets hot as the fuel burns

screen

fuel under test (The bottle is weighed before and after the experiment to see how much has been used.)

1 The bottle containing the fuel is weighed before and after the experiment. In this way, the mass of fuel that has burned can be calculated.

2 The temperature rise in the water is measured. The amount of heat energy needed to raise the temperature of 1 gram of water by 1°C is 4.2 joules, so the total amount of heat energy absorbed by the water can be calculated.

3 The screens and insulation cut down the heat lost to the air, but not completely. Some heat is also absorbed by the glass, but this can be allowed for in more complicated calculations.

4 The heat energy of burning (combustion) of the fuel is calculated in kJ per mole of fuel burned.

Here are some heat of combustion values for different fuels.

Fuel	Heat energy (kJ per mole)
Domestic gas (methane)	890
Bottled gas (butane)	2877
Ethanol (used in Brazil instead of petrol)	1367
Petrol	5513

Energy in food

In your body, fuels such as fats and sugars are used to provide warmth and energy. Energy values for foods can be measured by burning them in oxygen in a special apparatus. The table shows some values.

Food	Heat energy (kJ per kg)
fish (cod)	2800
potatoes	3500
beef	8000
bread	10 000
sucrose (sugar)	15 500
butter	30 000

Food gives you energy, but if that energy is not all used up, it will be stored as fat.

Figure 6
You use about 20 kJ min^{-1} walking briskly. How far could you walk using this packet as the 'fuel'?

An average serving of two Weetabix (37.5 g) provides at least one sixth of the daily recommended requirements for the average adult of the vitamins listed and iron.

TYPICAL NUTRITIONAL COMPOSITION

	Per 100 g			Per 100 g
Fat	2.0 g		Dietary Fibre	12.9 g
Protein	10.5 g		Vitamins:	
Available			Niacin	10.0 mg
Carbohydrate	66.8 g		Riboflavin (B_2)	1.0 mg
Energy	1400 kJ		Thiamin (B_1)	0.7 mg
	335 kcal		Iron	6.0 mg

Key points

- Fuels are chemicals which burn to produce heat and other forms of energy.
- Heat is given out in exothermic reactions and taken in in endothermic reactions.
- Heat of combustion may be measured by heating up water.
- Energy changes may be shown on an energy diagram.
- Food provides energy for our bodies.

Quick questions

1 Give two reasons why a space rocket must take an oxygen supply with it into space.

2 When a piece of magnesium burns in air, heat is given out. Draw an energy diagram for the reaction and write the symbol for the energy change.

3 Here are some food energy values.

Food	Energy in kJ per 100 g
bran flakes	1375
orange juice	192 (per 100 cm^3)
chewy bar	2080 (a bar = 150 g)
crispbread	1300 (each biscuit = 10 g)
cup soup	665 (per cup)
cheese	1660

A girl aged 16 needs about 9600 kJ each day to be healthy. Would the following slimmer's menu give her sufficient energy?

Breakfast 100 g of cereal
  250 cm^3 orange juice
Snack 1 chewy bar

Lunch 3 crispbreads and 150 g of cheese
Supper 1 cup soup and 2 crispbreads.

5.4 *Nuclear reactions*

A fair test

Much of the electricity in Britain is produced by nuclear power stations. Some people think the advantages of nuclear power (cheap electricity and no burnt fossil fuels to pollute the atmosphere) are outweighed by the disadvantages (the difficulty of disposing of nuclear waste and the danger of a nuclear explosion).

Some organisations, such as Greenpeace and Friends of the Earth say that the use of nuclear reactors exposes us to unnecessary amounts of radiation.

They say that statistics show children living near nuclear power stations and places where nuclear waste is processed, have a greater chance of contracting leukaemia (a blood disease which can be started by radiation) than others living elsewhere. They also say that radioactive waste put into the sea from nuclear reprocessing plants finds its way back into the environment, and that even the transportation of nuclear fuel by road or rail is dangerous because an accident might cause a leak of radiation.

The Central Electricity Generating Board (CEGB) reply that the amount of waste put into the sea is closely controlled. They say that the amount of radiation an individual absorbs from a power station is less than that absorbed from X-rays, from watching television or even the natural radiation from certain rocks in some parts of the country.

Often it is hard to decide which 'side' is right because of the difficulty of collecting enough information and then also interpreting it correctly. Sometimes, though, it is possible to test whether an opinion is supported by the facts or not.

The CEGB wanted to prove just how safe the transportation of nuclear fuel was by rail in giant steel containers called flasks. The flasks, each one weighing 48 tonnes, are carried on specially-made railway wagons hauled by diesel locomotives. First of all a flask was dropped from a great height, then crashed it into a railway tunnel wall at speed. Finally it was engulfed in flames. Each time, it was unharmed.

Next, in Summer 1984, they crashed a train at full speed into a nuclear flask. 450 tonnes of diesel locomotive, pulling three carriages was accelerated to a speed of 160 kilometres per hour and rammed into the flask, which was sitting stranded in the middle of the track. Five seconds later, when everything had stopped moving, there was one demolished locomotive and a completely unharmed nuclear flask.

This test gave very convincing evidence about the safety of the flasks in a similar kind of accident. However, not all differences in opinion on the safety of nuclear fuels are so easy to test objectively. Devising a fair test can often be difficult.

Figure 1
Can you think of a more drastic way of testing this nuclear flask? Does it prove flasks are always safe?

Nuclear fission

Some elements like uranium have atoms that are very big and unstable. When they are bombarded by neutrons, they split into smaller atoms, getting very hot in the process. At the same time, they eject more neutrons. This is called **nuclear fission**. Even in a small piece of uranium, there are many millions of atoms, so that many neutrons are ejected which each split more atoms, and each of these atoms gives out heat and more neutrons. Each time one uranium atom is split, three times as many atoms are split next time. This is called a **chain reaction**.

Figure 2
Nuclear fission.

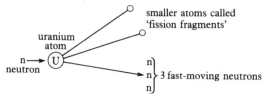

Figure 3
Each time a uranium atom is split, three more neutrons are released to split more uranium atoms. How many neutrons will there be after ten uranium atoms have split?

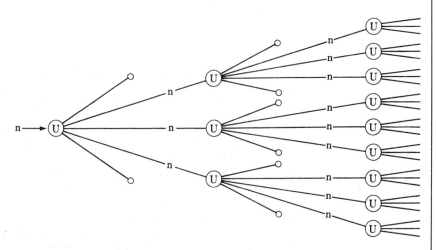

If a chain reaction is uncontrolled, the enormous release of heat turns into a nuclear explosion. In a **nuclear reactor**, the chain reaction is slowed down so that its heat can be used safely.

Nuclear fission can be written as an equation just like any other reaction. Here is one example.

$$^{235}_{92}U \quad + \quad n \quad \longrightarrow \quad ^{90}_{36}Kr \quad + \quad ^{142}_{56}Ba + 3n$$

uranium   neutron   smaller   more neutrons
atom   atoms   to split more
                                            uranium atoms

Nuclear reactors

In non-nuclear power stations, coal or oil is burned to make heat. The heat is then used to turn water into steam. The steam rotates a **turbine** connected to a **generator**. The generator makes electricity.

In a nuclear reactor, the heat from nuclear fission is used to turn water into steam. Figure 4 shows the inside of a nuclear reactor.

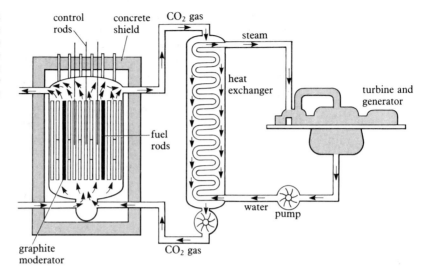

Figure 4
Nuclear fission produces heat. The heat can be used to make electricity in the same way as in an oil or coal fired power station, using a turbine and generator.

Each part of the reactor has a different function.

The **fuel rods** are magnesium alloy tubes that contain pellets of uranium 'fuel'.

The **moderator**, made of graphite, slows down the neutrons that come from the splitting atoms of uranium. Slow moving neutrons are better for fission reactions than fast moving ones.

The **control rods**, made of boron, absorb neutrons. They can be pushed in or pulled out of the core to control the rate of the chain reaction, or even to shut it down altogether.

The **concrete shield** keeps radiation in.

The carbon dioxide gas carries heat out of the reactor core to the **heat exchanger**. Here, the heat boils water to make steam for the turbine and generator. The gas is radioactive and must not escape.

In more advanced reactors, naturally occurring uranium is not used, but an **enriched** form is made instead where the proportion of isotope ^{235}U is increased to about 3%; ^{235}U atoms undergo fission more readily. In the **pressurised water reactor**, water at 150 times atmospheric pressure circulates around the core. This high pressure allows the water to be above 100°C without boiling. It acts as a moderator and also carries heat to the heat exchanger.

Nuclear reactors in Britain make about 13% of the total electricity and the CEGB plans to increase this to about 20% by the end of the 1990s.

Radioactivity

Some atoms have unstable structures and break apart easily. They are called **radioisotopes**. When they break apart, they give out energy or small particles which we call **radiations**. There are three sorts of radiation:

1 **α-particles** (α-radiation) are helium ions that come from the nuclei of some atoms when they break apart, or decay. They do not penetrate coverings such as gloves and overalls, but are dangerous if swallowed in food or breathed in dust.

2 **β-particles** (β-radiation) are very fast moving electrons that come from the nuclei of decaying atoms. They are more penetrating than α-particles, but can be stopped by thin sheets of aluminium foil.

3 **γ-radiation** is a form of electromagnetic radiation like X-rays, but far more penetrating. Several centimetres of lead or concrete are needed to stop this radiation.

Some radioisotopes occur naturally, but in very small amounts. The radiations that they give out is called **background radiation**. Other radioisotopes are man-made from atom bombs and radioactive waste from nuclear reactors. All forms of radiation are dangerous because they damage cells in the body. If a large amount of radiation is absorbed, so many cells are damaged that the victim dies within days or weeks. Smaller doses can be just as serious because the damaged cells can turn into tumours and cause cancer, often many years later. Radiation can also be used in the treatment of cancer, where the radiation is focused very accurately on the cancerous cells you want to kill.

The length of time that a radioisotope lasts is measured by its **half life**. The half life of a radioisotope is the time taken for its level of radiation to fall to half of its existing value. ^{137}Cs, one of the radioisotopes formed in the Chernobyl disaster has a half life of 33 years. As each 33 years go by, the radiation falls to half the previous amount. Many thousand years will have to pass before all the final radiation disappears.

Scientists measure the amount of radiation absorbed in **Sieverts**. A dose of 10 Sieverts or more absorbed in a few minutes would be fatal, but this would only happen in a very serious accident. Background radiation amounts to about 0.001 Sieverts per year and a hospital X-ray gives you a dose of about 0.0001 Sieverts. Workers in the nuclear industry absorb on average about 0.005 Sieverts per year. This is considered safe although different countries set different safety levels and opinions about safety levels differ.

Key points

- Some atoms, called radioisotopes, are unstable and split into smaller atoms, releasing radiation.
- The uncontrolled splitting of uranium atoms is called a chain reaction.
- Controlled nuclear fission can be used in nuclear reactors to produce heat for the generation of electricity.
- There are three forms of radiation, α, β and γ.

Quick questions

1 Explain how a very fast chain reaction builds up in a piece of uranium.

2 In a nuclear reactor, explain the job of (a) the concrete shield (b) the moderator (c) the control rods (d) the heat exchanger.

5.5 *Electrochemistry*

Sparks

This chapter is about the chemical reactions that take place when electricity flows through compounds. Gases like air do not conduct electricity normally. A gap of a few millimetres is quite sufficient to act as an insulator between two pieces of metal carrying electricity — unless the voltage is high, and then a spark jumps across the gap!

Figure 1
This giant spark causes chemical reactions in the air.

Big sparks, like lightning, cause chemical reactions to happen in the air. Oxygen is turned into ozone, and nitrogen and oxygen combine to make oxides of nitrogen. You can sometimes detect the bitter smell of these gases in the air if there has been a lightning strike nearby. The formation of oxides of nitrogen in this way is a very important part of the **nitrogen cycle** (see Section 7.5).

Conductors and non-conductors

Solid substances that allow electricity to flow through them are called **conductors**. All metals are good conductors of electricity.

Substances that do not allow electricity to flow through them are called **non-conductors** or **insulators**. All non-metal substances such as glass, plastic, wood and rubber are insulators. There is however, one very important exception. Carbon, in the form of graphite, is non-metal, but a very good conductor of electricity.

Electrolytes and non-electrolytes

> Liquids that will conduct an electric current are called **electrolytes**. Those that do not conduct are called **non-electrolytes**.

Electrolytes are substances that contain ions which are free to move about, so they are ionic compounds that are molten or are dissolved in water. Covalent compounds are non-electrolytes. Figure 2 shows the apparatus that could be used to see if a liquid is an electrolyte.

Figure 2
This apparatus could be used to test liquids to see if they are electrolytes.

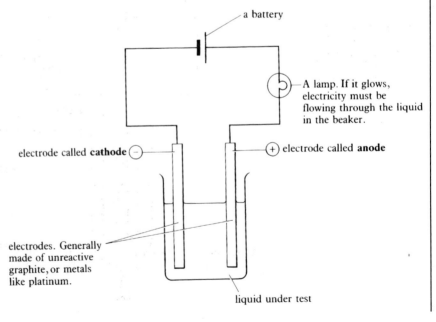

a battery

A lamp. If it glows, electricity must be flowing through the liquid in the beaker.

electrode called **cathode** (−)

(+) electrode called **anode**

electrodes. Generally made of unreactive graphite, or metals like platinum.

liquid under test

Type of substance	Electrolyte	Non-electrolyte
solutions of acids and alkalis in water	√	
solutions of salts in water	√	
molten salts	√	
ethanol petrol molten wax molten glass oil		√ √ √ √ √

Electrolysis

When electricity flows through a piece of metal, the metal is unchanged. When electricity flows through an electrolyte, a chemical reaction takes place and the chemical is decomposed. This reaction is called **electrolysis**.

The electrolysis of molten substances

The products formed when electrolysis takes place in a molten electrolyte are easy to predict from the electrolyte's name. (See Figure 3.) As the current flows, red fumes of bromine gas are produced at the positive electrode (the anode). At the end of the experiment, when the molten lead bromide is tipped out, a small piece of molten lead is left under the negative electrode (the cathode).

Figure 3
The electrolysis of molten lead bromide. What would you expect to see happen to the bulb if the liquid was not heated?

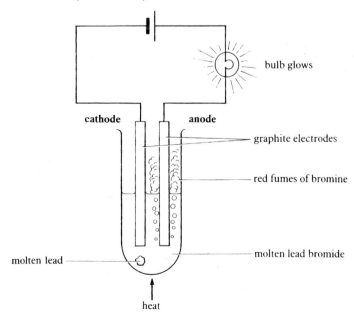

bulb glows

cathode   anode

graphite electrodes

red fumes of bromine

molten lead   molten lead bromide

heat

The table shows some more examples. Notice that the metal is always formed at the cathode.

Name of **molten** electrolyte	Product at anode	Product at cathode
lead(II) bromide	bromine	lead
sodium chloride	chlorine	sodium
aluminium oxide	oxygen	aluminium
potassium iodide	iodine	potassium
magnesium chloride	chlorine	magnesium

The process of electrolysis in more detail

When electrolysis takes place in an electrolyte, the ions move to the electrodes. Look at Figure 4. The negatively charged bromide ions are attracted to the positive electrode (the anode) and are changed into molecules of bromine gas.

$$2Br^- \longrightarrow Br_2(g) + 2 \text{ electrons}$$

Because they are attracted to the anode, the bromide ions are called **anions**.

The positively charged lead ions are attracted towards the negatively charged electrode (the cathode) and change into molten lead.

$$Pb^{2+} + 2 \text{ electrons} \longrightarrow Pb(l)$$

Because they are attracted to the cathode, lead ions are called **cations**.

During the electrolysis, electrons are given up by the anions and taken in by the cations. This movement of electrons from one electrode to the other means that electricity can flow around the circuit.

Figure 4
The electrolysis of molten lead bromide in more detail. Which ion is the cation?

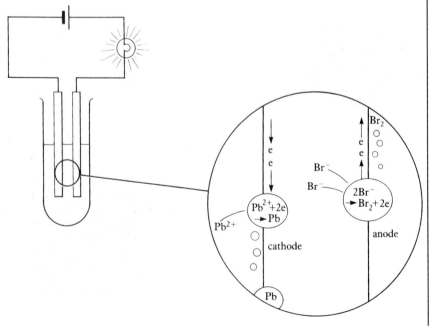

The positive and negative ions in ionic compounds are sometimes called by their electrolysis names of **cations** and **anions**. Negative ions are called anions and positive ions cations.

Name of ionic compound	Anion	Cation
lead(II) bromide	bromide ion	lead ion
sodium chloride	chloride ion	sodium ion
aluminium oxide	oxide ion	aluminium ion
potassium iodide	iodide ion	potassium ion

The electrolysis of substances dissolved in water

Solutions of substances in water are called aqueous solutions. Their electrolysis is more complicated because the water often takes part in the reaction too.

When water has substances dissolved in it, some of the water molecules split into ions.

$$H_2O \longrightarrow H^+ + OH^-$$
$$\text{hydrogen ion} \quad \text{hydroxide ion}$$

The electrolysis of dilute sulphuric acid

Look at Figure 5. Two gases are produced. At the anode, oxygen is given off and at the cathode, hydrogen is formed. The volume of the hydrogen is twice that of the oxygen.

Figure 5
The electrolysis of dilute sulphuric acid. Hydrogen and oxygen are produced in a volume ratio of 2:1. Why is a bulb included in the apparatus?

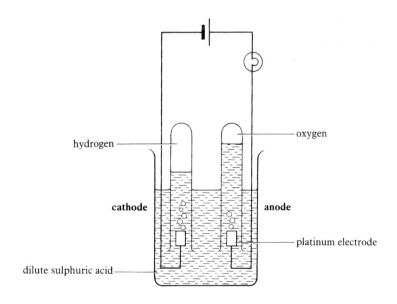

hydrogen

oxygen

cathode   anode

platinum electrode

dilute sulphuric acid

The electrolysis of dilute sulphuric acid in more detail

The ions present in dilute sulphuric acid are

H^+ and OH^- from the water
H^+ and SO_4^{2-} from the acid

There is only one cation (H^+). This is attracted to the cathode and changes into hydrogen gas.

$$2H^+ + 2 \text{ electrons} \longrightarrow H_2(g)$$

There are two anions (OH^- and SO_4^{2-}), but only the hydroxide ions take part in the electrolysis, changing into oxygen at the anode.

$$4OH^- \longrightarrow O_2(g) + H_2O + 4 \text{ electrons}$$

If the two equations are written so that the number of electrons in each reaction is the same, you can see why the volume of hydrogen is twice that of the oxygen.

$$4H^+ + 4e \longrightarrow 2H_2(g) : 4OH^- \longrightarrow O_2(g) + 2H_2O + 4e$$

same number of electrons

twice the volume
of gas

The electrolysis of copper sulphate solution

Figure 6 shows what happens when a solution of copper(II) sulphate is electrolysed. Oxygen gas is given off at the anode and the cathode becomes coated with a layer of pink copper.

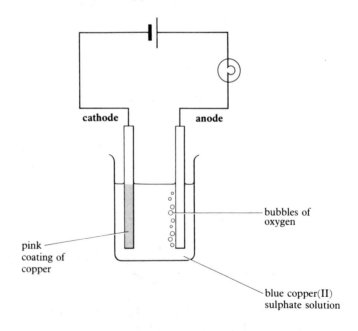

Figure 6
The electrolysis of copper(II) sulphate soultion. What do you think will happen to the colour of the liquid if the electrolysis is allowed to go on for some time?

cathode   anode

bubbles of oxygen

pink coating of copper

blue copper(II) sulphate solution

In more detail, this is what happens to the ions. There are two cations, H^+ from the water and Cu^{2+} from the copper(II) sulphate. Only the copper ions take part in the electrolysis.

$$Cu^{2+} + 2\ electrons \longrightarrow Cu(s)$$

There are two anions, OH^- from the water and SO_4^{2-} from the copper(II) sulphate. Only the OH^- ions take part in the electrolysis.

$$4OH^- \longrightarrow O_2(g) + 2H_2O + 4\ electrons$$

Predicting the products of electrolysis of aqueous solutions

Some anions take part in electrolysis much more easily than others.

$\left.\begin{array}{l} SO_4^{2-} \\ NO_3^- \end{array}\right\}$ almost never take part

$\left.\begin{array}{l} I^- \\ Br^- \\ Cl^- \end{array}\right.$ take part if the solution is concentrated

OH^- nearly always takes part

You can see that oxygen is nearly always formed, but if a concentrated solution of a salt is used, such as sodium chloride, chlorine will be formed as well.

With cations, the **reactivity** of the metal ion is important (see section 8.3). The less reactive the metal, the more likely it is to be formed during electrolysis.

K   **most reactive**
Na
Mg   hardly ever formed
Al
Zn
Fe   formed if the solution
Pb   is concentrated
H
Cu   most easily formed
Ag
Au   **least reactive**.

Although it is not a metal hydrogen must be in the list too. Because copper is less reactive than hydrogen, it is formed instead of hydrogen when copper sulphate solution is electrolysed.

Electroplating

The steel paper clip in Figure 7 is being nickel plated. The anode is made of nickel. As the current flows, the nickel dissolves to form nickel ions and electrons go into the circuit.

$$Ni(s) \longrightarrow Ni^{2+}(aq) + 2e$$

The paper clip is the cathode. Nickel ions from the electrolyte turn into nickel metal on its surface.

$$Ni^{2+}(aq) + 2e \longrightarrow Ni(s)$$

Figure 7
*When nickel plating a
paper clip, the clip must
be made the cathode in
the circuit.*

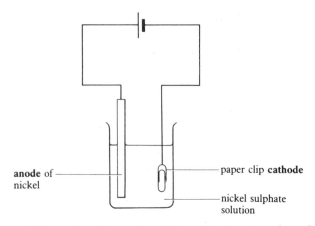

anode of —
nickel

— paper clip **cathode**

— nickel sulphate
solution

When electroplating, the object to be electroplated must replace the
cathode. The anode is generally made of the metal being plated and the
electrolyte must contain ions of the metal being plated.

Figure 8 shows parts of taps being chromium plated. So that the
chromium remains firmly on the steel underneath and doesn't flake off,
the tap must first of all be very clean and have all grease and dirt removed.
First it is copper plated and then nickel plated before the chromium plate
is put on. This gives a good 'finish' to the final stage when the chromium
is plated on. The tap is made the cathode in the electroplating bath and the
electrolyte for chromium plating consists of chromic acid and sulphuric
acid. Positive chromium ions are attracted to the cathode where they gain
electrons and change into chromium.

Figure 8
*Why is chromium used to
electroplate things like
the taps in this
illustration? Why is zinc
not used instead?*

Cells

During the electrolysis of an electrolyte, electricity makes a chemical reaction happen. In a cell the reverse is true. A chemical reaction takes place causing electrons from the chemicals to flow around the circuit making an electric current. A cell is commonly called a battery, although a battery is really a group of cells connected together. A car battery really **is** a battery, a torch battery is really a cell!

Two of the most important types of cell used in everyday things are the zinc carbon 'battery' and the nickel cadmium 'battery'.

The zinc carbon dry cell

This sort of cell is used in toys, radios, torches, portable electric shavers, personal hi-fi and many other things. Figure 9 shows what it looks like inside. The two electrodes in this cell are the zinc case and the manganese(IV) oxide powder. The electrolyte (not really 'dry' but soaked into paper) is ammonium chloride solution.

When the cell is connected into a circuit, the zinc reacts with the electrolyte, releasing electrons.

$$Zn(s) \longrightarrow Zn^{2+}(aq) + 2e$$

At the same time the manganese(IV) oxide is reacting

$$2MnO_2(s) + 2H_2O(l) + 2e \longrightarrow Mn_2O_3(s) + 2OH^-(aq)$$

This reaction carries on until the chemicals are used up. No matter how big the cell is and how much chemical it contains, the voltage of the cell is always 1.5V. 9V batteries, like the radio battery in Figure 10, contain the equivalent of 6 cells.

Figure 9
When the chemicals in this cell are used up, it stops making electricity.

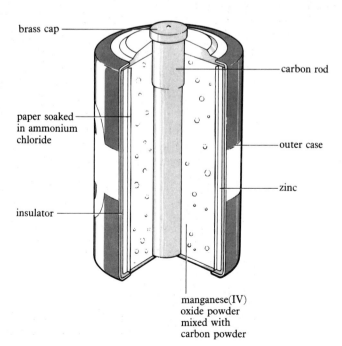

brass cap

carbon rod

paper soaked in ammonium chloride

outer case

zinc

insulator

manganese(IV) oxide powder mixed with carbon powder

The nickel cadmium cell

This contains sheets of nickel foil and cadmium foil, separated by paper soaked in potassium hydroxide solution. The sheets are wound up like a Swiss Roll and put inside a case. Both metals dissolve in the potassium hydroxide as the cell is used, but the chemical reaction is reversible and the cell can be used again. The nickel cadmium cell is a rechargeable cell. It is very useful for radios and tape recorders that get through batteries very quickly.

Figure 10 *This battery is small and compact to fit inside a radio. It contains smaller amounts of chemicals because the radio uses only a small current.*

Figure 11 *In rechargeable batteries of this sort, the chemical reaction which creates electricity can be reversed by passing electricity back into the battery. How much electricity will be needed to re-charge the battery?*

Key points

- Solids that conduct electricity are called conductors; liquids are called electrolytes.
- When electricity flows through an electrolyte, a chemical reaction takes place and this process is called electrolysis.
- Positive ions are called cations and negative ions are called anions.
- When aqueous solutions are electrolysed, the ions from the water often take part in the electrolysis.
- Cells are chemicals that react to produce electricity.
- If an object is coated with metal during electrolysis, it is said to be electroplated.

Quick questions

1 List three examples of each of the following: conductors, insulators, electrolytes, non-electrolytes.

2 Name the anions and cations in molten sodium bromide and magnesium iodide, and in dilute hydrochloric acid.

3 Draw and label the apparatus you would need to copper plate a key.

4 What products would you expect to get at the anode and cathode when a concentrated solution of potassium iodide is electrolysed.

5 Why would magnesium not be a suitable material for making electrodes for electrolysis?

6 For each of the following electrolytes, say what the products of electrolysis will be at the anode and cathode, if the electrodes are made of graphite.

   (a) dilute hydrochloric acid   (d) calcium bromide solution
   (b) sodium chloride solution   (e) silver nitrate solution
   (c) potassium iodide solution

Questions

1 Certain metals and some of their compounds are useful catalysts. For example, manganese(IV) oxide is a catalyst for the decomposition of hydrogen peroxide solution.

   (a) The volume of gas produced in this reaction was measured using the apparatus below.

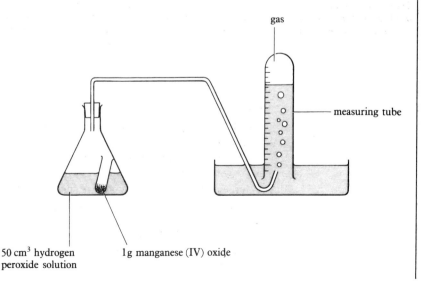

gas

measuring tube

50 cm³ hydrogen peroxide solution   1g manganese (IV) oxide

The results of the investigations are shown on the graph.

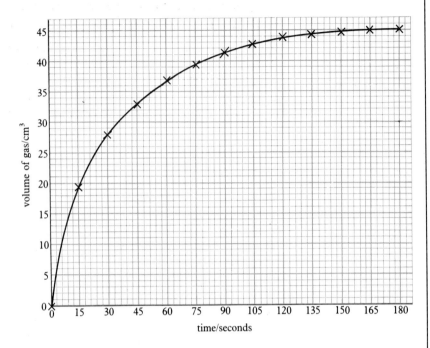

(i) What was the maximum volume of gas produced by the 50 cm^3 of hydrogen peroxide solution?

(ii) Name the gas produced by the decomposition of hydrogen peroxide.

(iii) At the start the rate of decomposition is fast. What causes the rate of decomposition to: slow down; and eventually stop?

(iv) At the end of the reaction the manganese(IV) oxide can be recovered by filtration. How could you show that the manganese(IV) oxide has acted as a catalyst?

(b) The experiment was repeated in exactly the same way except that this time the catalyst used was powdered iron. The results of this investigation were recorded in the table below.

Time/seconds	0	30	60	90	120	150	180
Volume of gas/cm^3	0	9	15	20	23.5	26.5	28.5

(i) Plot these points and draw the graph for these results on the same grid used for part (a).

(ii) What would be the maximum volume of gas produced by the 50 cm^3 of hydrogen peroxide using iron as the catalyst?

(iii) Which of these two catalysts is best for the decomposition of hydrogen peroxide? Give a reason for your choice.

(iv) Use the *collision theory* to explain why the speed of this reaction also depends on the temperature of the hydrogen peroxide.

SEG

2 1 g of a catalyst was added to 50 cm^3 of hydrogen peroxide. The volume of oxygen produced was noted every 10 seconds.

Time/seconds	0	10	20	30	40	50	60	70	80	90	100
Volume of oxygen/cm^3	0	19	35	35	54	58	59	60	60	60	60

(a)   (i) Plot these results on a graph like the one below.
(ii) Draw the best curve possible for these results.

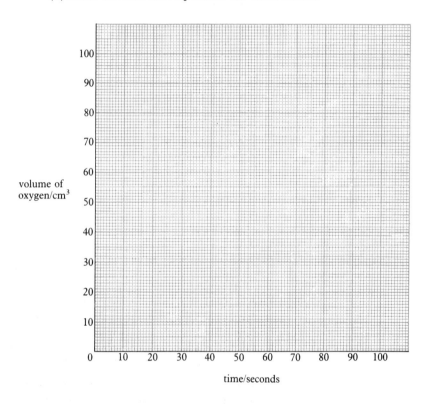

(b)   (i) Which one of the results appears to be an error?
(ii) When was the decomposition of the hydrogen peroxide fastest?
(iii) At what time did the reaction stop?
(iv) Give a reason why the reaction stops.

(c) Write an equation for this reaction.
(d) Draw a diagram of a suitable apparatus for this investigation.
(e) What effect does doubling the amount of catalyst have on the rate of this reaction? Give a reason for your answer.

SEG

* **3** The label on a bottle containing a brand of "Health Drink" reads as follows:

> *Ingredients: Glucose, vitamin C, flavouring essence (including caffeine), fruit acid, lactic acid, preservative, carbon dioxide.*
> *Carbohydrate content (as monosaccharide) 19.3 g per 100 cm³.*
> *Energy content 340 kJ per 100 cm³.*

(a) What colour would a piece of universal indicator paper become when dipped into the drink?

To check the "energy content", 10 cm³ of the health drink were heated on a water bath to evaporate all the water in the drink.

(b) Why was the health drink sample heated on a water bath instead of directly over a bunsen burner?

The solid residue from the evaporation was placed in a crucible and the apparatus shown below was set up.

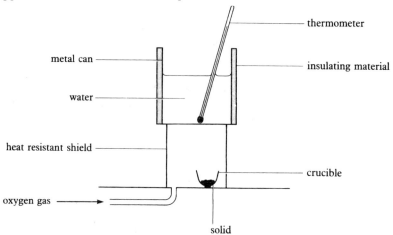

500 cm³ of water were placed in the metal can and its initial temperature recorded. The supply of oxygen was turned on and the solid ignited. The temperature of the water was recorded again once all the solid had burned.

Readings   Initial temperature of water = 19.6°C
                 Final temperature of water = 34.6°C

(c) Why was a metal can used to hold the water rather than a glass beaker?

(d) Calculate the amount of heat (in kJ) absorbed by the water in the can. (4.2 J of heat is needed to raise the temperature of 1 g of water by 1°C.)

(e) (i) Use your answer from **(d)** to estimate the 'energy content' of 100 cm³ of the health drink.
  (ii) Give **one** reason, other than heat losses to the air, why the 'energy content' measured by the experiment is different from that quoted on the label.

(f) The Heat of Combustion of glucose ($C_6H_{12}O_6$) is -3200 kJ/mol.
   (i) Write a balanced equation for the complete combustion of glucose.
   (ii) Draw an enthalpy level diagram to show the change in heat energy when glucose burns.

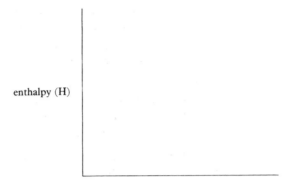

(g) Why is it useful for energy content values to be printed on many foods?

SEG

★ **4** A student used the following apparatus to pass an electric current first through sodium chloride solution and then through copper(II) chloride solution.

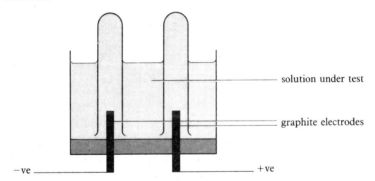

(a) In both cases the same mixture of gases was collected at the positive electrode. Name each of the two gases in the mixture from the results of the tests below.
   (i) Name the gas which relights a glowing splint.
   (ii) Name the gas which bleaches litmus paper.

(b) The product at the negative electrode was different in each case. Name the product formed at the negative electrode when using
   (i) sodium chloride solution;
   (ii) copper(II) chloride solution.

(c) Which particles carry the electric current through
   (i) graphite;
   (ii) sodium chloride solution?

SEG

* **5** The diagram below shows the electrolysis of sodium chloride solution.

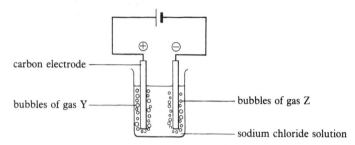

carbon electrode

bubbles of gas Y

bubbles of gas Z

sodium chloride solution

(a) Name the gases Y and Z.

(b) If a few drops of universal indicator are added to the solution before electrolysis starts, the indicator is green. As electrolysis happens, the indicator gradually turns blue.
  (i) What is the pH of the solution when the indicator is green?
  (ii) Explain why the electrolysis causes the indicator to go blue.
  (iii) Name a compound produced on an industrial scale by electrolysis of sodium chloride solution.

(c) In the electrolysis of copper(II) chloride solution the electron transfer at the negative electrode is shown by the following equation.

$$Cu^{2+}(aq) + 2e^- \longrightarrow Cu(s)$$

  (i) What does (aq) stand for?
  (ii) What would you expect to **see** at the negative electrode during this electrolysis?

(d) The table gives information about three substances A, B and C when they are solid and when they are molten.

	Solid substance		Molten substance		
Substance	Appearance of solid	Does the solid conduct electricity?	Does the melt conduct electricity?	Product at + electrode	Product at − electrode
A	white solid	no	yes	bromine	lead metal
B	yellow solid	no	no	(does not conduct)	
C	grey solid	yes	yes	none	none

  (i) Suggest possible identities for substances A and B.
  (ii) What type of bonding does solid B have?
  (iii) What type of bonding does solid C have?
  (iv) When the melted substance A conducts electricity what particles are carrying the current?

(e) (i) Predict the products of electrolysis of an aqueous solution of aluminium sulphate using inert electrodes. Explain how you arrive at your answer.
  (ii) Why is cryolite added to aluminium oxide during the electrolytic manufacture of aluminium?

SEG

6 Acids, alkalis and salts

6.1 Acids and alkalis

Fruity indications

Stewed blackcurrants are red, but when evaporated milk is poured over them, they go blue. The red chemical in the blackcurrants is an **indicator**. It goes one colour in acids (the fruit) and another colour in alkalis (the milk).

A problem to solve

Many other fruits and vegetables are indicators. Make a collection and see which fruit or vegetable is the most colourful indicator.

1 Put a piece of the fruit or vegetable into a few cm^3 of ethanol in a test tube, and gently prod with a glass rod to extract the colour. If the colour doesn't come out easily, put the test tube into a beaker of hot water to warm it up. (Figure 1.)

Figure 1
Why is the test tube of ethanol heated in a beaker of hot water instead of directly in the flame?

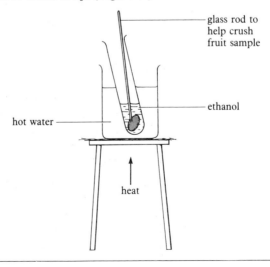

> **2** Centrifuge the test tube so that the clear coloured solution is separated from the remains of the fruit.
>
> **3** Put drops of your indicator solution into dilute hydrochloric acid and dilute sodium hydroxide solution (an alkali) and see what colour it turns. Record your results as shown below:
>
> | Fruit or vegetable | Colour in acid | Colour in alkali |
> |---|---|---|
> | blackcurrant raspberry rhubarb beetroot carrot | | |

Laboratory indicators

Blackberries and beetroot make good indicators, but their colours fade and liquids containing sugar grow mould. The indicator generally used in laboratories is **litmus** which is extracted from lichens which grow on the bark of trees. The litmus is dissolved in ethanol and used as a solution, or soaked into paper. Two other types of indicator are **methyl orange** and **phenolphthalein**.

Indicator	Colour in acid	Colour in alkali
litmus	red	blue
methyl orange	orange	yellow
phenolphthalein	colourless	pink

Laboratory acids

The table below shows acids commonly used in chemistry laboratories. They are used dissolved in water and are called dilute acids.

Acid	Formula	Ions present in the acid	
hydrochloric acid	HCl	H^+	Cl^-
nitric acid	HNO_3	H^+	NO_3^-
sulphuric acid	H_2SO_4	$2H^+$	SO_4^{2-}

All acids have one thing in common when they are dissolved in water. They form **hydrogen ions**, H^+. It is the hydrogen ions that make indicators change colour.

Laboratory alkalis

Alkalis that are commonly used in the laboratory are shown below.

Alkali	Formula	Ions present in the alkali	
sodium hydroxide	NaOH	Na^+	OH^-
potassium hydroxide	KOH	K^+	OH^-
calcium hydroxide	$Ca(OH)_2$	Ca^{2+}	$2OH^-$

All alkalis have one thing in common. They all form **hydroxide ions**, OH^- in solution. It is these ions that make indicators change colour.

Figure 2
Hydrogen ions in acids make litmus go red. Hydroxide ions in alkalis make litmus go blue.

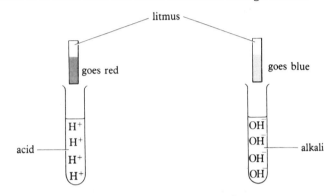

Acids and alkalis by other names

Many commonly used substances are acids or alkalis. Sometimes the acids can be tasted, as with lemons and vinegar. **Never taste anything in the laboratory!** Sometimes the alkalis can be smelled or detected by their soapy feel, like household ammonia and spray oven cleaner.

Name of substance	Chemical name	Acid	Alkali
vinegar	ethanoic acid	√	
lemon juice	citric acid	√	
rhubarb	ethandioic acid	√	
grape juice	tannic acid	√	
kettle descaler	aminosulphonic acid	√	
rust treatment	phosphoric acid	√	
bath scale remover	methanoic acid	√	
household ammonia	ammonia solution		√
spray oven cleaner	sodium hydroxide		√
Milk of Magnesia	magnesium hydroxide		√

pH numbers

Acids and alkalis in the laboratory are generally dissolved in water. The more water added, the more **dilute** the solution becomes. **Concentrated** solutions contain little or no water.

The **strength** of an acid depends upon the number of hydrogen ions present in its solution. Strong acids exist almost completely as their ions. Weak acids are only very slightly ionised in solution. Similarly the strength of an alkali depends upon how many hydroxide ions it has in solution.

To enable chemists to tell how strong an acid or alkali is, each different strength is given a **pH number**. The pH number of an acid or alkali can be measured with **Universal Indicator**. Universal Indicator is a mixture of indicators chosen so that it goes a different colour for most of the pH values. Universal Indicator can be used as a liquid, or soaked into paper, like litmus.

When added to an acid or alkali, the colour turned by the Universal Indicator can be compared with a chart like the one in Figure 3.

Figure 3
Each strength of acid or alkali has a pH number. Universal Indicator has a different colour for each pH number.

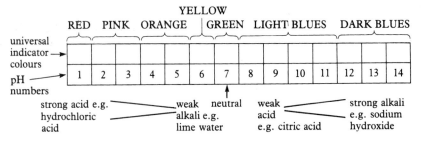

A problem to solve
--

A problem to solve

Stinging nettles sting! Dock leaves help reduce the pain of the sting when they are rubbed in, but why is this? One theory suggests that nettle sting contains an acid which is injected into the skin by the barbs on the nettle.

In the next section, you can read that alkalis neutralise acids. Could it be that dock leaves contain an alkali?

Devise an experiment to test this theory. Here is how you might start:
1 Extract some of the sting from the nettles by crushing them onto filter paper, or better still, by warming them with ethanol.
2 Test the extract with litmus or Universal Indicator.
3 ...

Key points

- Litmus is an indicator. Indicators go different colours in acids and alkalis.
- Acids are compounds that produce hydrogen ions when dissolved in water.
- Alkalis are compounds that produce hydroxide ions when dissolved in water.
- The strength of an acid or alkali is measured by its pH number. Universal indicator goes a different colour for each pH number.

Quick questions

1 Divide the following compounds into acids and alkalis by looking at their formulae:

$HNSO_3$ H_3PO_4 NH_4OH $Ba(OH)_2$ $HCOOH$ $LiOH$

2 Water which drains off limestone hills is weakly alkaline. What would you use to measure the level of alkalinity? Describe the results you might get.

3 Suggest a reason why stewed blackcurrants go blue when evaporated milk is added to them.

6.2 *Reactions of acids and alkalis*

Household cleaners

Figure 1
Household cleaning agents contain weak alkalis to dissolve grease.

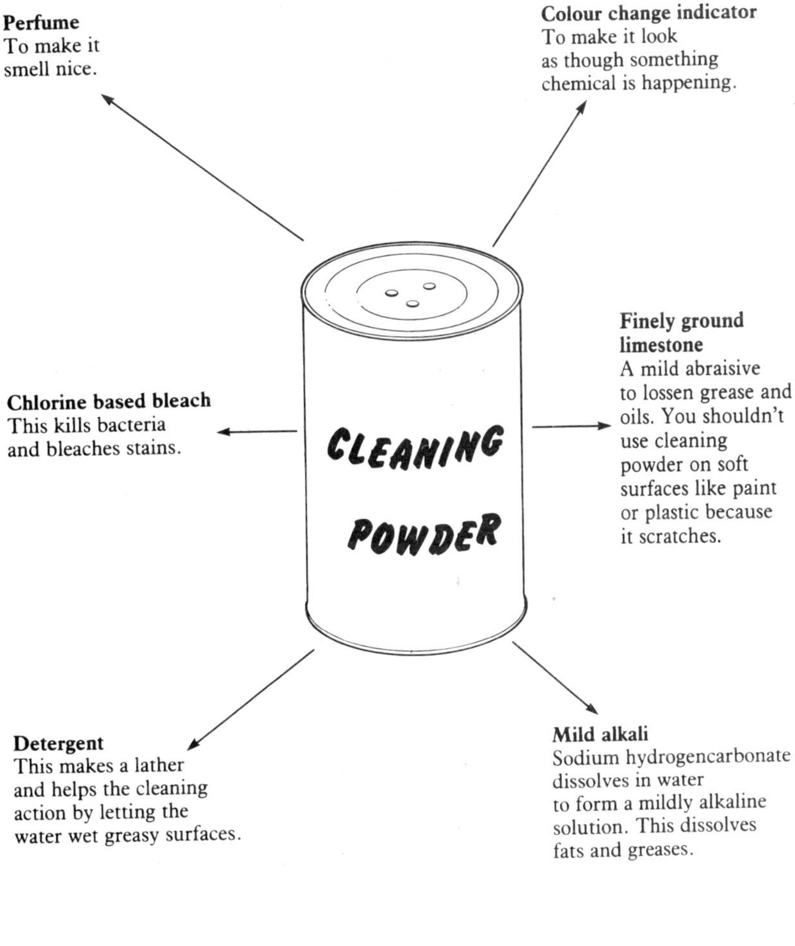

Perfume
To make it smell nice.

Colour change indicator
To make it look as though something chemical is happening.

Chlorine based bleach
This kills bacteria and bleaches stains.

Finely ground limestone
A mild abraisive to lossen grease and oils. You shouldn't use cleaning powder on soft surfaces like paint or plastic because it scratches.

CLEANING POWDER

Detergent
This makes a lather and helps the cleaning action by letting the water wet greasy surfaces.

Mild alkali
Sodium hydrogencarbonate dissolves in water to form a mildly alkaline solution. This dissolves fats and greases.

Fats and greases are dissolved by alkalis to make soap. This process is called **hydrolysis**. If you get some alkali on your fingers they feel soapy — and that is just what it is. The alkali has hydrolysed some of the fats in the tips of your fingers to make soap. Wash it off quickly! Alkalis such as sodium hydroxide are used to dissolve grease and fat on the inside of ovens. Household cleaning agents use much weaker alkalis.

Acids and metals

Most acids react with metals to form the gas **hydrogen** and leave a solution of a compound called a **salt**.

$$\text{metal} + \text{acid} \longrightarrow \text{a salt} + \text{hydrogen}$$

Here are two examples:

1 magnesium + hydrochloric acid $\longrightarrow$ magnesium chloride + hydrogen

$$Mg(s) + 2HCl(aq) \longrightarrow MgCl_2(aq) + H_2(g)$$

2 zinc + sulphuric acid $\longrightarrow$ zinc sulphate + hydrogen

$$Zn(s) + H_2SO_4(aq) \longrightarrow ZnSO_4(aq) + H_2(g)$$

Different reactivities

Figure 2 shows four different metals in dilute hydrochloric acid. The magnesium reacts very vigorously indeed and has dissolved in a few seconds. The zinc takes more time to react and is less fizzy. The tin hardly reacts at all and only produces a few bubbles. The copper does not react at all with the acid.

Figure 2
Different metals have different reactivities. Which of these metals is the most reactive?

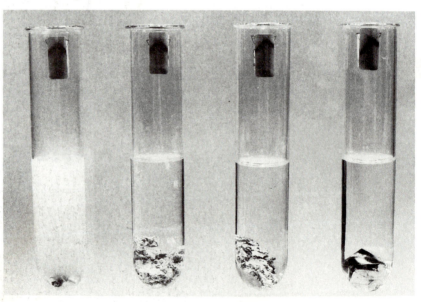

Mg   Zn   Sn   Cu

All metals can be put into a table called the **activity series**, depending upon how vigorously they react with acids, and also with other chemicals such as oxygen and water.

Figure 3
The activity series. What is the one non-metal in the series? Why do you think it is included?

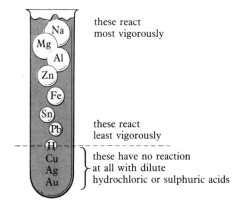

these react
most vigorously

these react
least vigorously

these have no reaction
at all with dilute
hydrochloric or sulphuric acids

Acids and carbonates

All compounds ending in 'carbonate' fizz in dilute acids and give off **carbon dioxide gas**. They leave behind a solution of a salt.

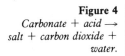

carbonate + acid ⟶ salt + carbon dioxide + water

Figure 4
*Carbonate + acid →
salt + carbon dioxide +
water.*

Here are two examples.

1   calcium carbonate $+$ hydrochloric acid $\longrightarrow$ calcium chloride $+$ carbon dioxide $+$ water

$$CaCO_3(s) + 2HCl(aq) \longrightarrow CaCl_2(aq) + CO_2(g) + H_2O(l)$$

2   copper(II) carbonate $+$ sulphuric acid $\longrightarrow$ copper(II) sulphate $+$ carbon dioxide $+$ water

$$CuCO_3(s) + H_2SO_4(aq) \longrightarrow CuSO_4(aq) + CO_2(g) + H_2O(l)$$

This reaction is used to make health salts, like Andrews, fizzy. (See Figure 1.) The acid in the health salts is **tartaric acid**. The carbonate is **sodium hydrogencarbonate**. Both are solids, but when water is added, they dissolve and react. The carbon dioxide formed makes the drink fizz.

Bases and acids

Bases are compounds which react with acids to form solutions of salts.

Base $+$ acid $\longrightarrow$ salt $+$ water

> Bases are oxides or hydroxides of metals.

CuO, Fe_2O_3, CaO, and ZnO are all bases. These bases all contain oxide ions.

$NaOH$, $Ca(OH)_2$, $Mg(OH)_2$, and $Pb(OH)_2$ are all bases too. These bases all contain hydroxide ions.

Bases which dissolve in water, such as sodium hydroxide, potassium hydroxide, calcium hydroxide and ammonia are called **alkalis**.

> Bases which dissolve in water are called alkalis.

When bases react with acids, the reaction is called **neutralisation**. These are examples of neutralisation reactions.

1   copper(II) oxide $+$ sulphuric acid $\longrightarrow$ copper(II) sulphate $+$ water

$$CuO(s) + H_2SO_4(aq) \longrightarrow CuSO_4(aq) + H_2O(l)$$

2   sodium hydroxide $+$ hydrochloric acid $\longrightarrow$ sodium chloride $+$ water

$$NaOH(aq) + HCl(aq) \longrightarrow NaCl(aq) + H_2O(l)$$

Ionic equations

Acids and bases are ionic substances. When they react, their ions react. For example, when copper(II) oxide reacts with sulphuric acid, the oxide ions in the base, and the hydrogen ions in the acid combine to make water.

$$O^{2-} + 2H^+ \longrightarrow H_2O$$

The other ions make the salt.

$$Cu^{2+}(aq) + SO_4{}^{2-}(aq) \longrightarrow CuSO_4(aq)$$

The same is true for sodium hydroxide and hydrochloric acid. The hydroxide ions from the alkali, and the hydrogen ions from the acid make water.

$$OH^- + H^+ \longrightarrow H_2O$$

The other ions make the salt.

$$Na^+(aq) + Cl^-(aq) \longrightarrow NaCl(aq)$$

These equations are called **ionic equations**. In any neutralisation reaction, the neutralisation part of the reaction is performed by the hydrogen ions, and the oxide or hydroxide ions from the base. The other ions are called **spectator ions** because they do nothing.

All neutralisation reactions can be written as

$$2H^+ + O^{2-} \longrightarrow H_2O \quad \text{or} \quad H^+ + OH^- \longrightarrow H_2O$$

Key points

- Acids react with most metals to form salts plus hydrogen; with bases to form salt plus water, and with carbonates to form salt plus carbon dioxide plus water.
- Alkalis are bases that dissolve in water.
- The reaction between an acid and a base to form a salt is called neutralisation.
- Ionic equations include only the ions which do any work in a reaction. The other ions are called spectator ions.

Quick questions

1 Write out and learn the formulae of the following acids and alkalis:
hydrochloric acid, sulphuric acid, nitric acid
sodium hydroxide, potassium hydroxide.

2 Complete these word equations:
(a) calcium + hydrochloric acid
(b) zinc carbonate + nitric acid
(c) aluminium oxide + sulphuric acid

3 Some magnesium is put into dilute sulphuric acid and allowed to fizz until all the magnesium dissolves. Does this mean that all the acid has been used up? Explain your answer and say how you would check it.

★ 4 For each of the following reactions, underline the spectator ions.
(a) $LiOH(aq) + HCl(aq) \longrightarrow LiCl(aq) + H_2O(l)$
(b) $ZnO(s) + H_2SO_4(aq) \longrightarrow ZnSO_4(aq) + H_2O(l)$
(c) $Ca(OH)_2(aq) + 2HNO_3(aq) \longrightarrow Ca(NO_3)_2(aq) + 2H_2O(l)$

5 Why do vinegar bottles have plastic or paper covers inside their metal caps?

6.3 *Salts*

Sea salt

Salt (sodium chloride) in Britain is mined from the ground. In hotter parts of the World, it is evaporated from sea water by the heat of the Sun. On the West coast of France, the land is flat and the weather is warm. Channels called 'etiers', bring water from the Atlantic Ocean to inland reservoirs from which it is fed into shallow pools cut in the ground. Here it slowly evaporates in the hot sunshine and fine crystals of sea salt form on the surface. These are scraped off with wooden rakes by men called 'Paludiers'.

Figure 1
What aspect of the weather controls the rate at which this salt is formed?

Salt from the sea is not just sodium chloride. Sea water contains magnesium, calcium and potassium ions as well as sulphate, bromide and fluoride. Both magnesium and bromine can be extracted from sea water in large quantities.

Some people say sea salt has much more flavour for cooking than ordinary salt. Why might this be? Why does sea salt cost more to produce than mined salt?

Naming salts

Salts are made from acids, so they are named after them.

>   salts made from **sulphuric acid** are called **sulphates**
>   salts made from **hydrochloric acid** are called **chlorides**
>   salts from **nitric acid** are called **nitrates**.

There are also carbonates, phosphates, chlorates, silicates and several others.

> Salts are made whenever the hydrogen ions in an acid are replaced by the metal ions from a base, a carbonate or a metal.

The name of the base, carbonate or metal gives the salt the other half of its name. For example

      calcium oxide + hydro**chloric** acid forms **calcium chloride**
      zinc + **sulphuric** acid forms **zinc sulphate**
      nickel carbonate + **nitric** acid forms **nickel nitrate**.

Soluble and insoluble salts

Some salts dissolve in water easily. They are said to be soluble. Other salts do not dissolve at all. They are said to be insoluble.

Soluble salts	Insoluble salts
nearly all sodium, potassium and ammonium salts	nearly all lead salts
all nitrates all sulphates ──────────────────────────────→	except barium, lead, calcium and mercury sulphates
all chlorides ──────────────────────────────→	lead, mercury and silver chlorides
except ←────────────────────────────── sodium, potassium, and ammonium carbonates	all carbonates

Making soluble salts

A metal, a base or a carbonate can be added to an acid until all its hydrogen ions are replaced by metal ions. The solution formed is then evaporated.

Making zinc sulphate from zinc and dilute sulphuric acid

1 25 cm^3 of dilute sulphuric acid is put into a beaker.

2 Small pieces of zinc are put into the acid. They fizz and dissolve.

      $Zn(s) + H_2SO_4(aq) \longrightarrow ZnSO_4(aq) + H_2(g)$

Why should this reaction be done in a fume cupboard, away from Bunsen flames?

3 More and more zinc is added until the reaction stops fizzing and some zinc is left over. Why is this important? What would it tell you has happened to the acid?

4 The mixture is filtered. What will be left in the filter paper?

5 The filtrate, zinc sulphate solution, is evaporated in an evaporating basin and then left to crystallise.

Figure 2
The sequence for making a soluble salt in the laboratory.

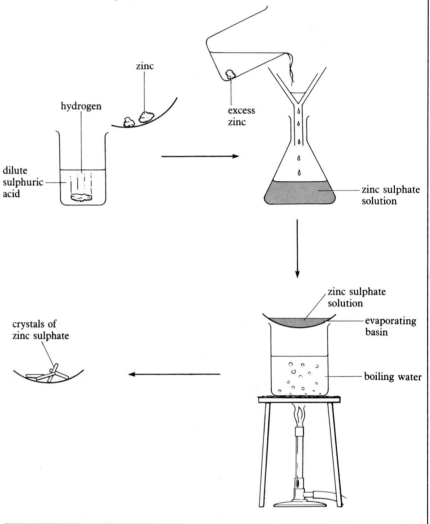

Figure 2 shows the sequence of events in making zinc sulphate from dilute sulphuric acid and zinc as a flow diagram. Zinc sulphate could also have been made using zinc carbonate instead of zinc. The same method and apparatus would be used, but the gas given off would be carbon dioxide and not hydrogen.

$$ZnCO_3(s) + H_2SO_4(aq) \longrightarrow ZnSO_4(aq) + CO_2(g) + H_2O(l)$$

Similarly, zinc oxide could have been used instead of zinc. No gas would be given off and the acid has to be warmed first.

Making salts from acids and alkalis

Making sodium chloride by titration

This sequence of events shown below are explained more fully in the numbered points below the diagram.

Figure 3
Titrating an acid with an alkali. Why should you always use a safety filler?

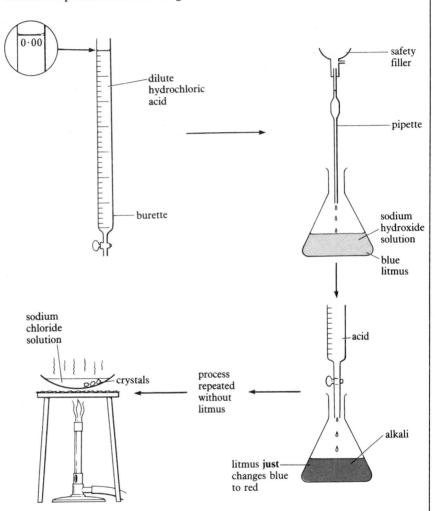

1 A burette is filled to the zero mark with dilute hydrochloric acid.

2 25 cm³ of dilute sodium hydroxide is put into a conical flask with a pipette. Why must a safety filler be used?

3 3 drops of litmus solution are put into the alkali. What colour will it turn?

4 Acid is run from the burette into the alkali. The flask is shaken. Why is this important? The moment the indicator changes to red (showing that the acid has just neutralised the alkali), the tap on the burette is closed and the volume of acid that has been added (called the **titre**) is noted.

5 The flask now contains a neutral solution of sodium chloride.

$$NaOH(aq) + HCl(aq) \longrightarrow NaCl(aq) + H_2O(l)$$

— but it is red!

6 The experiment is now repeated, using the same volume of alkali and the same titre of acid, but without the indicator. Why isn't the indicator needed this time?

7 Finally, the solution is evaporated to give white, cube-shaped crystals of sodium chloride.

Making insoluble salts

Salts which do not dissolve in water have to be made by the process of **ionic precipitation**. Precipitation is the formation of a solid when two solutions are mixed together.

Example 1

To make barium sulphate, a solution containing barium ions has to be added to a solution containing sulphate ions.

$$Ba^{2+}(aq) \quad + SO_4{}^{2-}(aq) \quad \longrightarrow \quad BaSO_4(s)$$

(This could	(This could
be barium	be dilute
chloride	sulphuric
solution)	acid)

Example 2

To make silver chloride, a solution containing silver ions must be added to a solution containing chloride ions.

$$Ag^+(aq) \quad + \quad Cl^-(aq) \quad \longrightarrow \quad AgCl(s)$$

(This could	(This could
be silver	be dilute
nitrate	hydrochloric
solution)	acid)

Making lead iodide by ionic precipitation

1 Insoluble lead iodide is made by mixing solutions containing lead ions and iodide ions.

$$Pb^{2+}(aq) + 2I^-(aq) \longrightarrow PbI_2(s)$$

Some ions have been missed out of this equation. What name is given to ions of this type?

2 About 20 cm³ of lead nitrate solution is put into a beaker. A similar volume of potassium iodide solution is added to it and the mixture is stirred. Why? A golden yellow precipitate of lead iodide forms.

3 The mixture is filtered and the lead iodide remains on the filter paper as the residue. What compound is in the filtrate?

4 The residue is washed by pouring distilled water over it while it is still in the filter paper. Why distilled water?

5 The paper containing the lead iodide is gently dried in the oven.

This sequence of events is shown in Figure 4 below.

Figure 4
The sequence for making an insoluble salt. Why is the residues washed before drying?

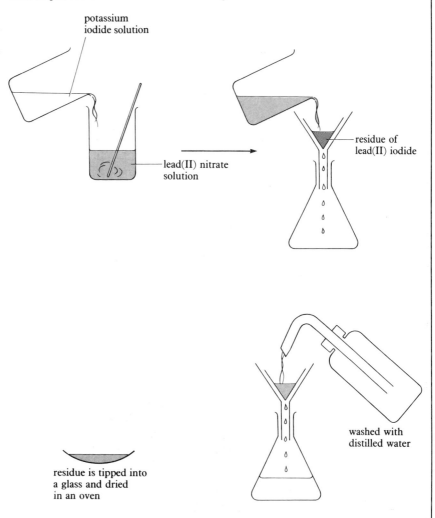

Water of crystallisation

Salts are ionic and when their solutions crystallise, the ions fit into a crystal lattice. Some salts fit molecules of water into the lattice as well. This is called **water of crystallisation**. If the water is removed by heating, the crystals often change in appearance. Not all salts have water of crystallisation, but here are some examples of salts that do.

Salts with water of crystallisation	Salts with no water of crystallisation
$CuSO_4.5H_2O$	NaCl
$CoCl_2.6H_2O$	KNO_3
$ZnSO_4.7H_2O$	$KClO_3$
$Na_2CO_3.10H_2O$	$KMnO_4$

The first two examples are important.

Example 1

Copper(II) sulphate-5-water; $CuSO_4.5H_2O$

When these blue crystals are heated, steam is given off and the crystals change to a white powder called **anhydrous copper(II) sulphate**.

$$CuSO_4.5H_2O(s) \longrightarrow CuSO_4(s) + 5H_2O(g)$$

If water is added to the anhydrous powder, it gets very hot and changes back into blue copper(II) sulphate solution.

$$CuSO_4(s) + water \longrightarrow CuSO_4(aq)$$

This is a **reversible** reaction. Anhydrous copper(II) sulphate can be used in this way to test for water.

Example 2

Cobalt chloride-6-water; $CoCl_2.6H_2O$

When these purple crystals are heated, they give off steam and change into a blue, anhydrous solid.

$$CoCl_2.6H_2O(s) \longrightarrow CoCl_2(s) + 6H_2O(g)$$

If water is added to the blue solid, it changes back to a pink-purple solution. Cobalt chloride solution can be soaked into filter paper and dried in an oven so that it goes blue. When the dried paper is dipped into water, it goes pink, so it can be used as a test for water.

Deliquescence

Some crystals of salts will absorb water from the atmosphere and turn into solution. These are said to be **deliquescent**. Examples of deliquescent salts are copper(II) nitrate and zinc chloride. Sodium hydroxide is also a very deliquescent solid. (See Figure 5.) To prevent deliquescent substances from deliquescing, they are kept in a **desiccator**. (See Figure 6.)

Efflorescence

Some crystals of salts give out some or all of their water of crystallisation. These salts are said to be **efflorescent**. An example is sodium carbonate-10-water, $Na_2CO_3.10H_2O$. When transparent crystals of this salt are left open to the air, they become white and powdery on the surface as they effloresce. (See Figure 7.)

$$Na_2CO_3.10H_2O(s) \longrightarrow Na_2CO_3.H_2O(s) + 9H_2O(g)$$

Figure 5
Which one of these dishes of sodium hydroxide pellets has been exposed to air?

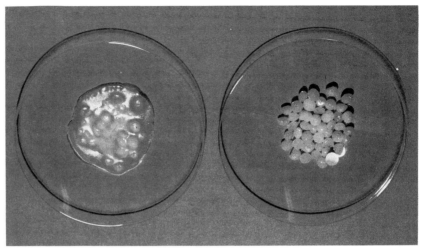

Figure 6
The lumps of calcium chloride in the bottom of the desiccator absorb moisture from the air so that things in the top may be kept dry.

Figure 7
Sodium carbonate crystals lose nearly all their water of crystallisation when they effloresce.

Key points

- Salts are named from the acid and base (or carbonate or metal) from which they are made.
- Insoluble salts are made by ionic precipitation.
- Some salts have water of crystallisation.
- Deliquescent salts take in water from the atmosphere.
- Efflorescent salts lose water of crystallisation to the air.

Quick questions

1 Name the salts made from these chemicals.
  (a) magnesium and hydrochloric acid
  (b) cobalt carbonate and sulphuric acid
  (c) iron(III) oxide and nitric acid
  (d) potassium hydroxide and hydrochloric acid

2 What methods would you use to make the following salts?
  Hint: first of all check if they are soluble or insoluble.

      sodium nitrate   mercury chloride
      calcium carbonate   aluminium chloride
      iron(II) sulphate   lead(II) sulphate

3 Explain why anhydrous copper(II) sulphate should be kept in a desiccator.

Questions

1 When answering this question, use the names of the compounds given here:

      carbon dioxide   sodium chloride
      copper(II) carbonate silver nitrate
      potassium hydroxide zinc oxide
      silver chloride   zinc sulphate

  (a) (i) Which one of the compounds will give off a gas when dilute hydrochloric acid is added to it?
     (ii) Give the name or formula of the gas formed.

  (b) (i) Which one of the compounds will dissolve in water forming a solution which will neutralise nitric acid?
     (ii) Write a word equation for the reaction which occurs.

  (c) (i) Name two compounds which, when dissolved in water and the solutions are mixed, will form a white precipitate of an insoluble salt.
     (ii) Give the name or formula of the salt formed.

(d) (i) Name a compound which is a white solid, is insoluble in water
and neutralises dilute sulphuric acid forming a soluble salt.
(ii) Give the name or formula of the salt formed.

NEA specimen

2 Complete the table below, which describes the preparation of some salts.

Reactants			Products		
magnesium oxide	+	$\longrightarrow$	magnesium sulphate	+	
	+	$\longrightarrow$	zinc chloride	+	hydrogen
	+ sodium sulphate	$\longrightarrow$	lead sulphate	+	

SEG

3 Magnesium sulphate crystals ($MgSO_4.7H_2O$) can be made by adding
excess magnesium oxide (MgO), which is insoluble in water, to dilute
sulphuric acid.

(a) Why is the magnesium oxide added in excess?

(b) The following apparatus could be used to separate the excess
magnesium oxide from the solution. Label the diagram.

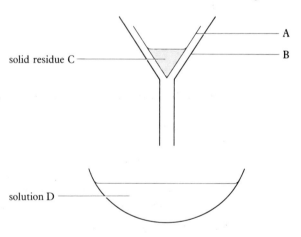

(c) Given the relative atomic masses: H = 1, Mg = 24, O = 16,
S = 32, calculate the relative formula mass of
(i) Magnesium oxide, MgO
(ii) Magnesium sulphate crystals, $MgSO_4.7H_2O$

(d) Use your answers in **(c)** to calculate the maximum mass of
magnesium sulphate crystals that could be obtained from 2.0 g of
magnesium oxide.

(e) Describe how you would obtain pure, dry crystals of magnesium
sulphate from magnesium sulphate solution.

LEAG

4 Medicines to treat an upset stomach often contain magnesium hydroxide. These medicines work by neutralizing some of the hydrochloric acid in the stomach.

(a) The pH of the acid in the stomach is about 1. What happens to this pH when a medicine to treat an upset stomach is taken?

(b) Why is sodium hydroxide **not** suitable for curing an upset stomach?

(c) Hydrochloric acid contains the ions H^+ and Cl^-.
Magnesium hydroxide contains the ions Mg^{2+} and OH^-.
  (i) Which one of these ions would cause a solution to be acidic?
  (ii) Which one of these ions neutralizes acidity?
  (iii) Name and give the formula of the salt formed from the reaction of hydrochloric acid and magnesium hydroxide.

SEG

7 Non-metals and their compounds

7.1 An introduction to non-metals

Carbon fibres

Carbon, in the form of **graphite**, is a brittle material. It breaks and snaps easily when a strain is put upon it. It is quite unlike a metal which can be bent and flexed and can bear great loads without giving way. The reason for this difference between non-metals like carbon and metals lies in their structure. All materials, no matter how smooth and polished they may seem have tiny cracks in their surfaces. In non-metals such as carbon, these cracks can spread quickly throughout the material and cause the non-metal to snap or shatter. The internal structure of metals does not allow this to happen.

However, things are quite different if carbon is made into tiny, whisker-like **fibres**. The surfaces of the whiskers are so smooth that cracks cannot develop, and if the fibres are embedded in a resin or plastic, the material becomes extremely strong. Carbon fibre mixtures like this can be fifty times stronger than steel, but lighter than aluminium. It is not surprising that this material is used in the aircraft industry. Blades in the compressors of jet engines and the huge blades on helicopters are being made of carbon fibre materials today. (See Figure 1.)

Carbon fibres conduct heat very efficiently. In Concorde, the brake linings are made of carbon fibre resins, instead of the more usual asbestos with steel, to conduct away the heat generated. Carbon fibres conduct the heat more efficiently and are much lighter too. The weight saved means eight more passengers can be carried!

Non-metals in the periodic table

In the periodic table non-metals are found on the right hand side. This is because of their electron structures. (See Figure 2.)

Figure 1　*These blades are made from carbon fibre material which is stronger and lighter than metal.*

1	2	3	4	5	6	7	8	9	10	11	12	13	14	15	16	17	18
H																	He
Li	Be											B	C	N	O	F	Ne
Na	Mg											Al	Si	P	S	Cl	Ar
K	Ca	Sc	Ti	V	Cr	Mn	Fe	Co	Ni	Cu	Sn	Ga	Ge	As	Se	Br ★	Kr
Rb	Sr	Y	Zr	Nb	Mo	Tc	Ru	Rh	Pd	Ag	Cd	In	Sn	Sb	Te	I	Xe
Cs	Ba	La	Hf	Ta	W	Re	Os	Ir	Pt	Au	Hg ★	Tl	Pb	Bi	Po	At	Rn
Fr	Ra	Ac															

- metals
- non-metal solids
- non-metal gases

★ *liquids at room temperature*

La	Ce	Pr	Nd	Pm	Sm	Eu	Gd	Tb	Dy	Ho	Er	Tm	Yb	Lu
Ac	Ph	Pa	U	Np	Pu	Am	Cm	Bk	Cf	Es	Fm	Md	No	Lw

Figure 2　*The position of the non-metals in the Periodic Table.*

Structure and appearance of the non-metals

The table gives some typical non-metals and their properties.

Element	Appearance	Structure
carbon (graphite)	black, brittle solid electrical conductor	giant structure (Figure 3)
carbon (diamond)	very hard, transparent non-conductor	giant structure (Figure 3)
sulphur	yellow, brittle solid non-conductor	ring of 8 atoms (Figure 4)
phosphorus	waxy yellow solid non-conductor	4-atom molecule (Figure 4)
iodine	black crystals non-conductor	2-atom molecule (Figure 4)
bromine	blood red liquid	2-atom molecule
hydrogen, chlorine, nitrogen, oxygen	gases	have 2-atom molecules
helium, neon	both gases	both exist as single atoms

Figure 3 *Carbon can exist as graphite or diamond. Both have continuous, giant structures.*

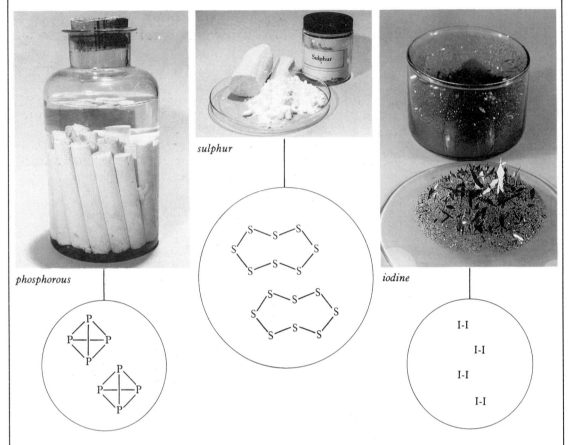

phosphorous

sulphur

iodine

Figure 4   *The formulae of phosphorus, sulphur and iodine should correctly be written as P_4, S_8 and I_2*

Key points

- Non-metals take up the right hand side of the periodic table.
- Solid non-metals are brittle and non-conductors of electricity, except for graphite which does conduct electricity.
- Solid non-metals exist as giant structures (graphite and diamond), or molecules with 2 or more atoms.
- Gas molecules all have 2 atoms except for the Noble gases, which exist as single atoms.

Quick questions

1 Why can't you
   (a) make electrical wires out of phosphorus
   (b) make a saucepan out of sulphur
   (c) make a bridge out of graphite?

★ 2 (a) Why doesn't neon form molecules?
   (b) Why doesn't helium burn?

7.2 *Air and its gases*

Aerosols

Substances which need to be dispensed in fine sprays, like hair spray, insect killer and perfume; or in foams, like shaving cream or oven cleaner, are often put into **aerosol** cans. When the button on top of the can is pressed, the **propellant gas** forces the liquid product up the dip-tube and into the nozzle, where it is forced out in tiny droplets or as a frothy foam. Without the propellant gas, the aerosol will not work.

Figure 1
Why won't this aerosol can work upside down?

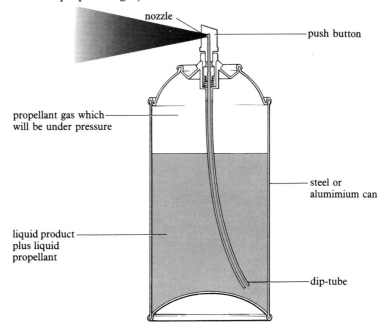

nozzle

push button

propellant gas which will be under pressure

steel or alumimium can

liquid product plus liquid propellant

dip-tube

Not just any old gas will do, however. It must be

- non-poisonous
- odourless
- non-flammable
- it mustn't react chemically with the product in any way
- it must be liquefied in the can so that a steady and lasting pressure can be produced as the aerosol is used.

Ordinary compressed air will not do. Dinitrogen oxide (nitrous oxide) is fine for things like cream and milk products, and carbon dioxide is used with car windscreen de-icers. Although they are flammable, hydrocarbons like butane work well with water based products like polishes.

However, the best propellants, developed over the past 50 years, are **chlorofluorocarbons** such as

dichlorofluoromethane	$CHCl_2F$
trichlorofluoromethane	CCl_3F
and dichlorotetrafluoroethane	$C_2Cl_2F_4$

But even in 1974, it was suspected by American scientists that chlorofluorocarbons released from aerosols were diffusing into the atmosphere and, in a series of chemical reactions, were decomposing the **ozone layer**. Ozone (O_3) exists at a low concentration (about 1 part of ozone to 100 000 parts of air) in a layer about 25 km above the surface of the Earth. The ozone layer is essential to us because it filters out much of the **ultraviolet light** coming from the sun. Too much ultraviolet light causes skin cancer. Recent measurements have shown that the ozone layer is being damaged more quickly than was originally supposed and that a hole has now appeared in the ozone layer over the Antarctic.

Aerosols are not the only culprits. Chlorofluorocarbons are used in refrigerators (and get released when the fridge is broken up) and are used to put the gas into polystyrene foam. Also, oxides of nitrogen from car exhausts and gases from the jet engines of high flying aeroplanes destroy ozone as well. Despite the convenience of deodorants in pressure cans and hamburgers in insulated polystyrene boxes, serious steps have been taken to reduce the amount of chlorofluorocarbons used. Countries in the EC have agreed to try to reduce the use of these compounds by 30%. Sweden and the USA have banned their use in aerosols altogether.

A lungful of air

When you take a deep breath of clean, unpolluted air, you take in the mixture shown in Figure 2. Surprisingly, the amount of each of these gases stays the same, despite all the things that are done in air.

Figure 2
A lungful of air. Which gas is the most plentiful?

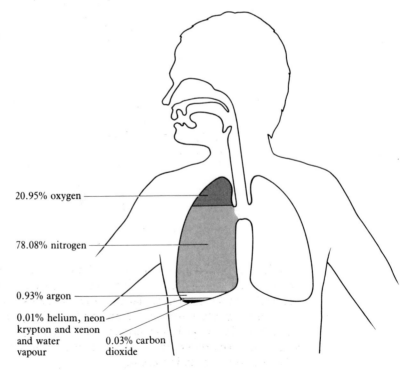

20.95% oxygen

78.08% nitrogen

0.93% argon

0.01% helium, neon krypton and xenon and water vapour

0.03% carbon dioxide

Figure 3
In your lungs, oxygen is transferred into the bloodstream so that it can be used in all the cells in the body. Carbon dioxide from the cells is passed back into the lungs from the blood.

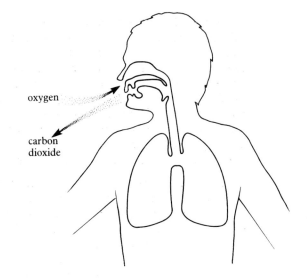

Figure 4
Green plants contain chlorophyll, a catalyst that enables them to convert water and carbon dioxide into sugar and oxygen, using the sun's energy.

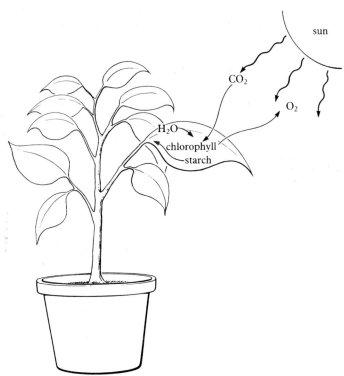

Respiration and burning (**combustion**) are processes which use up oxygen and produce carbon dioxide. Natural processes like decomposition of vegetable matter and fermentation also produce carbon dioxide. On the other hand, **photosynthesis**, the process by which plants produce food, uses up carbon dioxide and puts oxygen into the atmosphere. These reactions all balance each other out — although in fact the amount of carbon dioxide is very slowly increasing.

Measuring the amount of oxygen in the air

1   When copper is heated in air, it combines with the oxygen, but does not react with the other gases. The copper turns into black copper(II) oxide.

$$2Cu(s) + O_2(g) \longrightarrow 2CuO(s)$$

Figure 5
The hot copper in the tube combines with the oxygen in the air. How could you test the apparatus before starting to see if it had a leak in it?

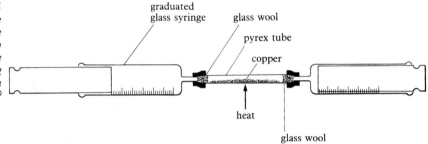

2   One of the syringes shown in Figure 5 is filled with 100 cm³ of air. The other syringe is empty. The pyrex tube contains small pieces of pink copper wire.

3   The copper is heated as the air is pushed from syringe to syringe over the copper.

4   As the copper reacts with the oxygen in the air, it changes from pink to black, showing that copper(II) oxide is being formed. The heat is slowly moved along the tube. How will you know when all the oxygen in the air has been used up?

5   The apparatus is allowed to cool down before the next measurement is made. Why is this? The volume of cold air left in the syringe is noted.

6   Results of this experiment might look like this:

Volume of air at start	=	100 cm³
Volume of air at end	=	79 cm³
Volume of O_2	=	21 cm³
Percentage of oxygen in the air	=	21%

Using air

Air is a mixture of gases. Thousands of tonnes of air are split up each day in a process called the **fractional distillation** of liquid air, in order to separate these gases for the uses listed below.

First of all the air is cleaned and cooled so that it condenses into liquid. The mixture of liquids (the liquid air) is then warmed so that each different gas boils as it reaches its own boiling point as the temperature rises. This process is fractional distillation.

Oxygen is used to make steel, for high temperature cutting and welding equipment, and for breathing apparatus at high altitude, deep sea, and in hospitals. Two million tonnes of oxygen are made and used in this way in the UK each year.

Nitrogen is used in liquid form to cool air for deep frozen food and in gaseous form for making ammonia and nitric acid. Some of it is transported in insulated tankers like the one shown in Figure 6.

Argon, a Noble gas that does not react with oxygen, is used as an inert atmosphere for welding certain metals and for filling electric light bulbs. (Figure 7). **Neon** glows red when an electric current is put through it and is used in lamps and advertisement signs. **Helium** is almost as light as hydrogen and doesn't burn so it is used in weather balloons.

Figure 6
This vacuum-flask on wheels contains liquid nitrogen. Why is nitrogen transported as a liquid rather than a gas? What would happen to the liquid nitrogen if the tank were punctured?

Figure 7 *Electric light bulbs are filled with argon. Why not leave air inside? Lights filled with neon glow red.*

Air pollution

Air pollution is a very serious problem in big industrial cities. San Francisco is in a valley, so warm air from the city containing car fumes and factory smoke becomes trapped by a layer of cold air. Unburnt hydrocarbons, oxides of nitrogen and carbon monoxide react in the sunlight to form ozone and other harmful chemicals. This mixture is called a **photochemical smog**. It damages people's health, stunts the growth of plants and trees and cuts out the sunlight. It only clears when cars stop running. Smogs of this sort occur wherever there is serious air pollution or where geographic features provide the right conditions.

Figure 8
San Francisco has smogs because warm air gets trapped underneath a layer of cold air.

In Great Britain there are no longer any smogs since laws were passed in the 1950's to prevent factories and houses putting large quantities of smoke into the air. However, there is still air pollution. Car engines put carbon monoxide and lead dust into the air. Carbon monoxide is a poisonous gas that prevents haemoglobin in the blood from absorbing oxygen. Large amounts of the gas in the blood can lead to heart failure. Lead is poisonous and even in small amounts is particularly dangerous to young children. The lead damages brain cells and the nervous system. The EC recommends that the lead content of petrol should now be only 0.15 g per litre of petrol and that all petrol should be lead-free by 1991.

It isn't only chemicals like these that cause problems. Carbon dioxide, which is made whenever any fossil fuel like coal or oil is burned, is slowly building up in the Earth's atmosphere. This gas absorbs the Sun's heat more than ordinary air and the Earth's surface is slowly heating up. This is known as the **greenhouse effect** and could have serious consequences in the future including the melting of the Polar ice caps.

- If the ice caps did melt, what effect would this have on low-lying parts of the World?

- If all forms of air pollution were banned tomorrow, what are some of the things you would have to do without in your everyday life?

Reactions with oxygen

When things burn, they combine with oxygen to form compounds called oxides. Substances burn more brightly in pure oxygen than in air because the nitrogen in the air slows down the burning process by reducing the concentration of the oxygen.

Burning elements in oxygen

1 Several Pyrex test tubes are filled with oxygen and corked.
2 A small sample of an element like sulphur is heated on a combustion spoon until it catches fire, and then it is lowered into the oxygen. (Figure 9.)
3 The sulphur burns with a bright blue flame, producing a misty gas. When it has finished burning, a piece of damp blue litmus is put into the gas. The litmus goes red, showing that the gas is acidic.
4 The experiment is repeated using samples of other elements.

Figure 9
Which form of litmus would be best when testing the product formed in this experiment — litmus paper or litmus solution?

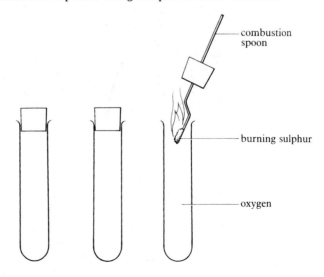

combustion spoon

burning sulphur

oxygen

The reactions of different elements with oxygen are given in detail in other sections, but here is a summary of the sorts of oxides obtained.

Non-metals such as sulphur, phosphorus and carbon burn in oxygen to form oxides that turn litmus red.

> Most oxides of non-metals are called acidic oxides.

Metals such as zinc, iron, lead and copper react with oxygen to form oxides which react with acids to form salts.

Metals such as sodium, potassium and calcium react with oxygen to form oxides which dissolve in water to form solutions of alkalis. They turn damp litmus blue.

> Most oxides of metals are called basic oxides. Some basic oxides dissolve in water to form alkaline solutions.

Oxidation

When an element or compound reacts with oxygen, it is **oxidised**.

For example,

$$C(s) + O_2(g) \longrightarrow CO_2(g)$$
$$CH_4(g) + 2O_2(g) \longrightarrow CO_2(g) + 2H_2O(g)$$

The carbon has been oxidised and turned into carbon dioxide. The methane has been oxidised and turned into carbon dioxide and steam. These are **oxidation** reactions and the oxygen is called the **oxidising agent**.

Key points

- Aerosol propellant gases damage the ozone layer.
- Respiration uses up oxygen. Photosynthesis uses carbon dioxide and produces oxygen.
- The air is a mixture of gases which can be separated by fractional distillation.
- The gases of air all have important uses.
- Air pollution is a very complicated problem that cannot be cured overnight.
- Elements burn in oxygen to form oxides: metal oxides are basic, non-metal oxides are acidic.
- Oxidation is the process of gaining oxygen.

Quick questions

1 How do the processes of respiration, combustion and photosynthesis balance each other out?

2 What volume of (a) oxygen (b) nitrogen (c) argon would you expect to find in 200 cm^3 of air?

3 Describe a source of air pollution in your neighbourhood. How could it be reduced or stopped?

4 Give one use for each of the components of the air.

★ 5 Describe an experiment, using oxygen, by which you could see if an element was a metal or a non-metal.

7.3 Hydrogen

The fuel of the future?

If the Earth's store of oil and gas is all used up, there will be no more petrol or diesel. They could be made from coal, but would be very expensive made in this way. Instead, hydrogen might be the new fuel of the future.

Hydrogen is plentiful. Most of the Earth's surface is covered by water and hydrogen may be made from this compound quite easily. Hydrogen burns with a hot, clean flame and produces only steam, so as a fuel, it would not pollute the air with carbon monoxide and unburnt hydrocarbons, as petrol does.

The problem with hydrogen is how to store it. If it is stored in steel cylinders at a pressure of 135 times atmospheric pressure, 99% of the weight is that of the steel cylinder: only 1% is the hydrogen! Cars burning hydrogen would have to carry a lot of extra weight around. However, there are ways to overcome this problem. Scientists in West Germany have found that tiny glass beads, less than one hundredth of a millimetre across can absorb hydrogen when under pressure, and release it again when gently heated. Glass is lighter than steel!

Another possible fuel is metal hydrides. Hydrogen can be made to react with certain metals such as iron, magnesium, nickel or palladium to make metal hydrides. These compounds absorb large quantities of hydrogen gas which is released again when the hydride is heated. The hydrogen could be released for example, by the heat from the car's engine as it goes along.

● Do cars of the future have to be powered by a fuel like hydrogen? Are there other ways of getting energy?

Industrial uses of hydrogen

In Great Britain, 100 million m^3 of hydrogen gas is made each year. Nearly half of this is used to make ammonia and nitric acid. The rest is used for making chemicals from oil, for plastics such as Nylon and for the food industry. Some hydrogen is a by-product of the electrolysis of sodium chloride, but most is made by the process of **steam reforming**.

In steam reforming, methane (Natural Gas) is mixed with steam and heated to a high temperature over a catalyst of nickel.

$$CH_4(g) + H_2O(g) \longrightarrow CO(g) + 3H_2(g)$$

The products of the reaction are carbon monoxide and hydrogen, which are later separated.

Hydrogen in the laboratory

Hydrogen is made when metals such as zinc and magnesium react with dilute acids like hydrochloric acid.

$$Zn(s) + 2HCl(aq) \longrightarrow + ZnCl_2(aq) + H_2(g)$$

Figure 1 shows the apparatus needed. Because hydrogen does not dissolve in water it can be collected by bubbling it through water into a gas jar.

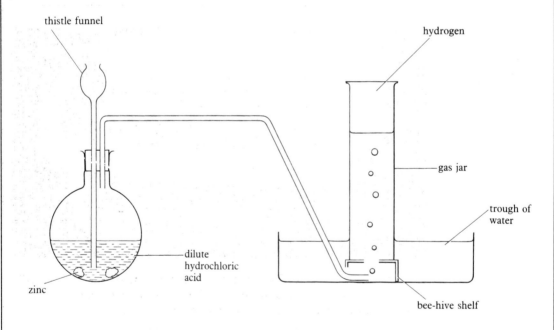

thistle funnel

hydrogen

gas jar

trough of water

dilute hydrochloric acid

zinc

bee-hive shelf

Figure 1   *The thistle funnel allows you to add more acid. Why must the end of the thistle funnel tube be below the surface of the acid?*

Margarine

More than a hundred years ago, Napoleon's soldiers spread it on their bread as part of their emergency rations. Napoleon III offered a prize for a cheap, energy-giving food, that would keep better than butter. His scientists came up with a mixture of animal fats, sour milk and salt — it was margarine.

Today, a variety of animal and vegetable fats and oils are used to make different sorts of margarines. Beef fat, and fish oils from herring, pilchards, sardines and anchovies, are blended with oils from soya beans, palm nuts, coconuts, maize, rapeseed and sunflowers. First the fats and oils are purified by mixing with an alkali, then washed and finally steamed to remove colour and smell. The fats are solidified by heating the oils with hydrogen over nickel catalysts in a process called **hydrogenation**. This changes the oils' structures, making them into solids. Finally the hydrogenated oils have **emulsifiers** (to keep them mixed up), colouring, flavouring, vitamins A and D, salt and skimmed milk added to make the final product.

Different types of margarine contain different proportions of oils and fats. For example, hard cooking margarines have a higher proportion of animal fats in them. Soft spreading margarines contain only vegetable oils.

Figure 2
What is this margarine made from? Do you think all the ingredients are necessary?

Hydrogen is the lightest of all elements. Because its molecules are so small and fast moving, it mixes, or diffuses, with other gases very quickly. This is why the 'squeaky pop' test for hydrogen works. (See Figure 3.)

Hydrogen also escapes from tiny holes more quickly than any other gas. Figure 4 shows two balloons which were blown up the day before. One balloon, the one still inflated, contains carbon dioxide. The other one containing hydrogen has gone down. This is because the hydrogen molecules are smaller than the carbon dioxide molecules and can escape through the tiny holes in the rubber more easily.

Figure 3
Hydrogen molecules move very quickly. As soon as the test tube is opened, some hydrogen escapes and air gets in. An explosive mixture is formed and the lighted splint makes it explode, or "pop".

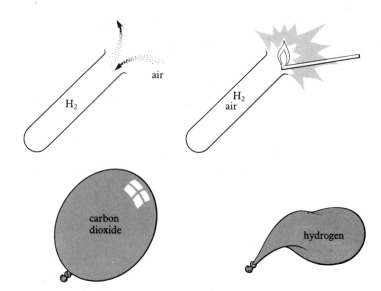

Figure 4
After a day, one balloon is still inflated, but the other has gone down. Which has the smaller molecules, carbon dioxide or hydrogen?

If hydrogen were ever to be used instead of Natural Gas as a fuel, the danger of leaks would be a problem. However, since hydrogen burns with a clean, blue flame to produce only water it would reduce pollution.

The combustion of hydrogen

1 Hydrogen from a cylinder is allowed to flow through a U-tube of calcium chloride (to remove any water from it) and then out of a jet. After a few seconds, the hydrogen is lit. (Figure 5.) Why isn't the gas lit immediately?

Figure 5
Calcium chloride dries the hydrogen. Why bother to do this? Will the cold water stay cold for long? How will cobalt chloride paper and a thermometer help you to find out if water is produced and not another chemical?

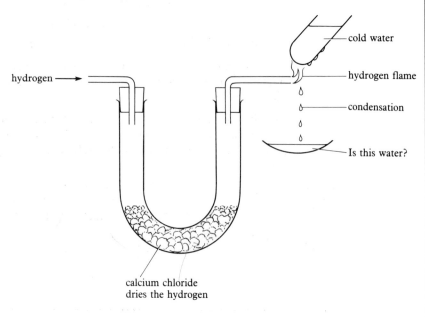

calcium chloride
dries the hydrogen

2 When hydrogen burns, only steam is formed.

$$2H_2(g) + O_2(g) \longrightarrow 2H_2O(g)$$

The steam condenses on the cold test tube and forms water.

How could you demonstrate that the liquid was water, and not some other liquid using (a) cobalt chloride paper (b) a thermometer?

If hydrogen is mixed with air and then ignited, a very fast burning reaction takes place. So much heat and steam is given out in a short space of time that there is an **explosion**, so hydrogen can be dangerous.

Reduction

Oxidation is the gain of oxygen. An element or compound is oxidised if it combines with oxygen. **Reduction** is the opposite of oxidation. A compound is reduced if it has oxygen removed from it. The substance removing the oxygen is called a **reducing agent**. Hydrogen is a good reducing agent. It will reduce compounds such as copper(II) oxide and lead(II) oxide.

Reduction of oxides

1 Small amounts of lead(II) oxide and copper(II) oxide are put into porcelain dishes inside a pyrex tube.

2 Hydrogen is piped in from one end (from a cylinder) and flows over the oxides and out of the other end. It is allowed to flow for a few seconds to push all of the air out before it is lit. (Figure 6.)

3 The oxides are heated and the hydrogen reduces the oxides.

$$PbO(s) + H_2(g) \longrightarrow Pb(s) + H_2O(g)$$
$$CuO(s) + H_2(g) \longrightarrow Cu(s) + H_2O(g)$$

Why doesn't the hydrogen burn inside the tube?

4 Lead(II) oxide is yellow. When it is reduced, it changes to silvery beads of lead. Copper(II) oxide is black. When it is reduced, it glows and changes to pink copper.

5 Finally, the apparatus is allowed to cool before the hydrogen is turned off and the lead and copper taken out. If air got in whilst they were still hot, they would oxidise back to lead(II) oxide and copper(II) oxide.

Figure 6
What precautions must be taken when lighting the hydrogen flame?

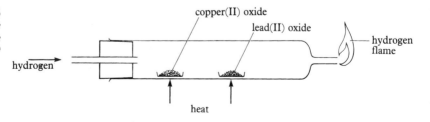

copper(II) oxide
lead(II) oxide
hydrogen flame
hydrogen
heat

Key points

- Hydrogen is a fuel. It burns with a hot, clean flame and causes very little pollution.
- Hydrogen is used in the production of margarines and in the manufacture of ammonia and nitric acid.
- Most hydrogen is manufactured by the steam reforming of methane.
- In the laboratory, hydrogen is made when an acid reacts with a metal.
- Hydrogen burns to form only steam.
- Hydrogen is a reducing agent. Reduction is loss of oxygen.

Quick questions

1 Explain why hydrogen makes a good fuel.

2 Which pair of gases will mix more quickly: air and carbon dioxide or air and hydrogen? Explain your answer.

3 Laboratories using hydrogen and other flammable gases, sometimes use electrical switches immersed in oil. Can you think why?

4 What will (a) ZnO (b) Fe_2O_3 (c) SnO_2 be turned into when they are reduced?

5 In the reaction:

$$CuO(s) + C(s) \longrightarrow Cu(s) + CO(g)$$

which substance is the reducing agent and which substance is being reduced?

7.4 *Carbon and limestone*

Diamonds

Diamonds are formed from pure carbon which has been changed by intense heat and pressure, deep in the Earth many millions of years ago. Diamond-containing rock, found in Russia, South America and South Africa, is crushed and washed and the diamonds are separated, sometimes using X-rays to detect them. More than 30 tonnes of diamond-containing rock is needed to yield 1 carat (0.2 g) of diamond. Apart from their beauty, diamonds are valued because they are so hard. Most are used in industry for cutting and drilling. The drill bit in Figure 1 has diamond embedded in it for extra hardness.

Some of the bigger and better-shaped diamonds are cut to be made into gem stones. Experts split the diamonds using stainless steel blades to obtain straight edges and flat surfaces. The surfaces are then polished with diamond powder to make them smooth. The diamonds are cut into shapes that trap light, reflecting it around inside the diamond so that it sparkles. The 'brilliant cut' diamond in Figure 2 has 58 faces or 'facets'.

The biggest diamond ever discovered was found in a South African mine in 1902. It weighed 3106 carats — more than half a kilogram!

Figure 1
When drilling through hard rock, only diamonds are hard enough.

Figure 2
This diamond sparkles because light is reflected back from the cut surfaces.

Allotropy

Carbon can exist in two quite different forms: **diamond** and **graphite**. Diamond is made when carbon is subjected to very high temperatures and pressures. Graphite occurs naturally as black crystals, and in such things as soot and charcoal. This property of existing in two solid forms is known as allotropy and diamond and graphite are called **allotropes**.

> When an element can exist in two or more different forms (without melting or boiling), it is said to have the property of allotropy.

The two allotropes of carbon are quite different because of their structures.

Figure 3

Diamond is a continuous structure where each carbon atom is joined to four other atoms in a tetrahedral arrangement. Graphite is made of sheets of carbon rings. The layers are stacked on top of each other.

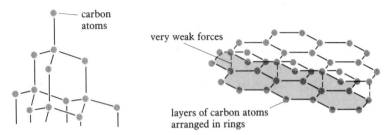

- In **diamond**, the carbon atoms are arranged in a tetrahedral structure.
- Diamond is the hardest known natural substance because of the strength and tetrahedral arrangement of the bonds between the carbon atoms.
- Diamond is an electrical insulator because all the carbon electrons are used to make the covalent bonds between the carbon atoms.
- In **graphite**, the carbon atoms are arranged in sheets of rings of six atoms. The sheets are piled on top of each other but there are no covalent bonds between the layers.
- Graphite is a soft, crumbly substance which feels greasy. It feels like this because the layers can easily slide over each other. Graphite is used as a lubricant.
- Graphite conducts electricity. Each carbon atom has four electrons in its outside shell but only three are used in bonding. There are electrons free to allow an electric current to flow.
- Graphite is black because the random arrangement of the layers does not allow light to pass through.

However, diamond and graphite are both pure carbon and burn in the same way (although diamond needs a higher temperature).

$$C(s) + O_2(g) \longrightarrow CO_2(g)$$

Moh's hardness scale

10 minerals are arranged in order of their hardness:

1 talc	6 feldspar
2 gypsum	7 quartz
3 calcite	8 topaz
4 fluorspar	9 corrundum
5 apatite	10 diamond

When a geologist wants to identify an unknown mineral, he or she scratches it with one of these ten stones. If the mineral can be scratched it is softer than the Moh stone being used. If not, then it is harder.

In this way, the geologist gets some idea of the identity of the unknown mineral from its hardness.

Limestone

Limestone is a sedimentary rock, consisting mainly of calcium carbonate, formed millions of years ago from sediments in shallow rivers and seas. It often contains fossils of tiny sea-shell animals. Limestone is quarried (Figure 4) and then used for building, steel making, or turned into lime for manufacturing sodium carbonate, cement, agriculture and water treatment and numerous other uses.

Figure 4
The crane in this picture gives an idea of how big the quarry is. The limestone rocks are broken up with explosives.

To make **lime** the crushed limestone is mixed with coke and heated in tall towers (kilns) with very hot air. As the coke burns, the limestone decomposes into calcium oxide or lime.

$$CaCO_3(s) \longrightarrow CaO(s) + CO_2(g)$$

Lime is used in agriculture for soils that are too acidic. It neutralises the acid in the soil and helps break up clay as well. Lime is also added to some water supplies in an attempt to neutralise water polluted by acid rain.

Cement is a mixture of lime and calcium sulphate. Powdered limestone is mixed with clay, coal dust and water. The mixture is heated in a big retort at 1500°C. The coal burns and the heat produced makes the limestone decompose into lime. After cooling and crushing, gypsum (calcium sulphate), is added before it is put into bags.

Mortar is made when cement is mixed with water and sand. When mortar dries, the lime and gypsum form a strong interlocking lattice with the sand and set hard. If small stones are added, the mixture is stronger still and is called **concrete**. Reinforced concrete, used for making bridges and building has steel mesh embedded in it for extra strength.

Heating limestone

If limestone is heated, it decomposes.

$$CaCO_3(s) \longrightarrow CaO(s) + CO_2(g)$$
calcium oxide

This reaction takes place at about 800°C. The calcium oxide (called **quicklime**) can be added to water to form calcium hydroxide (called **slaked lime**).

$$CaO(s) + H_2O(l) \longrightarrow Ca(OH)_2(aq)$$
calcium hydroxide

Calcium hydroxide is alkaline and a solution of calcium hydroxide is called **lime water**.

Limewater

When carbon dioxide gas is bubbled through limewater, a neutralisation reaction takes place:

$$CO_2 + Ca(OH)_2(aq) \longrightarrow CaCO_3(s) + H_2O$$

carbon dioxide + limewater $\longrightarrow$ calcium carbonate

acidic gas   alkali   insoluble salt

The insoluble calcium carbonate appears as a white suspension. The limewater goes 'milky'.

> When carbon dioxide is bubbled through limewater, the limewater goes milky.

Make your own limewater

Marble is another form of limestone made millions of years ago when limestone deposits in the earth's crust were compressed. Marble is harder than limestone and can be broken into small pieces without crumbling.

1 Put a small marble chip on the end of a loop of wire and heat it in the hottest part of a Bunsen flame for several minutes so that it decomposes to calcium oxide.

$$CaCO_3(s) \longrightarrow CaO(s) + CO_2(g)$$

2 Let the chip cool down completely. Notice how the outside of the chip has become crumbly as it has turned into calcium oxide.

3 Put the chip into 20 cm^3 of water in a beaker and stir thoroughly. There should be a hiss and the water should get hot as the calcium oxide dissolves to form calcium hydroxide solution.

$$CaO(s) + H_2O(1) \longrightarrow Ca(OH)_2(aq)$$

4 Filter the mixture to remove any undecomposed limestone and keep the filtrate.

5 Using a straw, gently blow through the filtrate. The filtrate is limewater and it should go milky.

This sequence of events is summarised below.

Figure 5
A marble chip is heated, cooled and reacted with water. Unreacted marble is filtered off. The filtrate (limewater) goes milky when you blow through it.

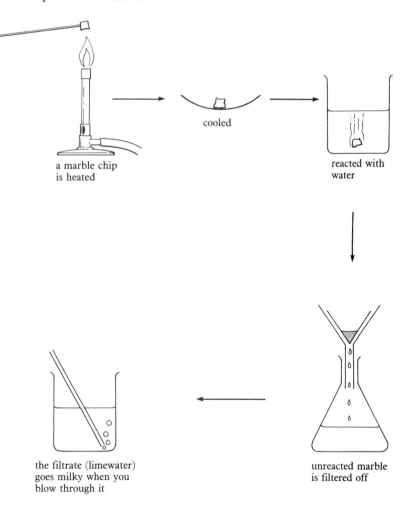

a marble chip
is heated

cooled

reacted with
water

unreacted marble
is filtered off

the filtrate (limewater)
goes milky when you
blow through it

● What gas in your breath makes the limewater milky?
● What could you do to the marble chip at the start of the experiment to reduce the time needed for heating?

Heating other carbonates

Most carbonates behave in the same way as calcium carbonate when they are heated. They decompose into an oxide and give off carbon dioxide.

carbonate ———→ oxide + carbon dioxide

Sodium and potassium carbonate, however, are too stable to decompose. They melt before they decompose.
Calcium and **magnesium** carbonates need high temperatures to decompose. A roaring Bunsen flame is necessary.
Lead and **copper** carbonates decompose much more easily at a lower temperature.

$$PbCO_3(s) \longrightarrow PbO(s) + CO_2(g)$$
white   yellow

$$CuCO_3(s) \longrightarrow CuO(s) + CO_2(s)$$
green   black

Carbonates and acids

All carbonates react with dilute acids in the same way. They all fizz and dissolve, giving off carbon dioxide and leaving a solution of a salt.

carbonate + acid ———→ a salt + carbon dioxide + water

For example

1 calcium carbonate + hydrochloric acid ———→ calcium chloride + carbon dioxide + water

$$CaCO_3(s) + 2HCl(aq) \longrightarrow CaCl_2(aq) + CO_2(g) + H_2O(l)$$

2 copper(II) carbonate + sulphuric acid ———→ copper(II) sulphate + carbon dioxide + water

$$CuCO_3(s) + H_2SO_4(aq) \longrightarrow CuSO_4(aq) + CO_2(g) + H_2O(l)$$

Carbon dioxide

Carbon dioxide is a heavy, colourless gas which dissolves in water slightly, at ordinary pressures.

It dissolves better at higher pressures and is used to put the fizz into fizzy drinks. The gas is bubbled under pressure through the drink before the can or bottle is sealed. When the container is opened, the pressure returns to normal and the gas 'undissolves' and bubbles out.

At −78°C, carbon dioxide changes directly from a gas to a solid. This process of missing out the liquid stage is called **sublimation**. Solid carbon dioxide is called '**dry ice**' and is used as a refrigerating material.

Carbon dioxide does not let things burn: it does not support combustion. Because of this it is used in fire extinguishers. The heavy gas forms a blanket over the flames, cutting off the oxygen supply.

Figure 6
Carbon dioxide is used on electrical fires. Why would it be dangerous to use water?

Key points

- Carbon has two allotropes called graphite and diamond.
- The difference between graphite and diamond is explained by their different structures.
- Limestone (calcium carbonate) is used for many purposes, including cement making and agriculture.
- Limewater, a solution of calcium hydroxide in water, turns milky when carbon dioxide is bubbled through it.
- Most carbonates decompose when heated to give carbon dioxide and an oxide.
- Carbonates fizz with acids and give off carbon dioxide.
- Carbon dioxide is used in fire extinguishers and dry ice.

Quick questions

1 What are allotropes?

2 What is it about their structures that makes diamond the strongest substance and graphite a lubricant?

3 Name one mineral that will be scratched by gypsum.

4 What will happen if iron(II) carbonate is put into dilute sulphuric acid? Name the products and write the equation.

7.5 *Nitrogen compounds*

Explosives

An explosion is a very fast combustion reaction. A burnable material is made to react with oxygen to form gases such as steam and carbon dioxide. If the reaction takes place quickly enough, a very large volume of hot, expanding gas is produced and a shock wave is created in the air. This is heard as an explosion.

One of the earliest explosives, known for more than a thousand years, is gun powder. If saltpetre (potassium nitrate), charcoal (graphite) and sulphur are mixed together in the correct proportions, the mixture burns very rapidly when ignited. The potassium nitrate decomposes to make oxygen, which oxidises the carbon to carbon dioxide. Large volumes of nitrogen are produced too. If the gun powder is put into a cardboard tube, the gases split the tube apart with an explosion.

High explosives started to be made in the mid 19th Century. Nitro-cellulose (gun cotton) was made from cotton and nitric acid, and nitroglycerine was made from glycerine and nitric acid. These substances are good explosives because all that is needed for combustion (carbon, nitrogen and oxygen atoms) is contained in one molecule, and the reaction is very rapid. But this also makes these molecules very unstable and difficult to handle. Nitroglycerine is a liquid which burns quite safely until its temperature reaches 220°C: then it explodes violently, and even if jolted, it can explode.

Perhaps the most famous name in the history of explosives is that of Alfred Nobel, a Swedish explosives manufacturer who lived from 1833 until 1896. Alfred discovered that if nitroglycerine was soaked into sticks of clay, it was quite safe to handle and could be set off only by a small explosion called a detonator. He called his new explosive, **Dynamite**.

Figure 1
The man in this old picture is making nitroglycerine in one of Nobel's factories. He no doubt worked long hours at this dangerous job. Why do you think his stool had only one leg?

Alfred Nobel died in 1896, and he left all his money to be invested so that the interest could be given as prizes to the people who had done the best work in Physics, Chemistry and Medicine, and for the best work in Peace and Literature.

The nitrogen cycle

Nitrogen plays a vital part in our lives because without nitrates, plants cannot live. Figure 2 shows how the nitrogen is recycled all the time in what is called the **nitrogen cycle**.

Figure 2
The Nitrogen Cycle.

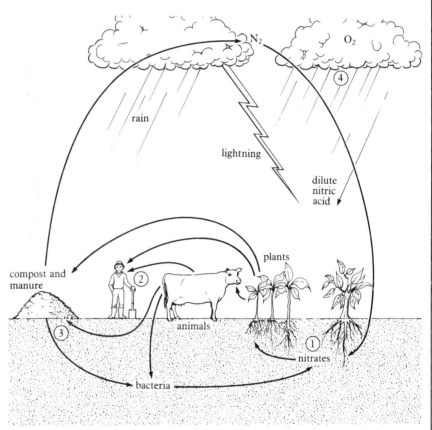

1 Nitrates are very soluble in water and are absorbed by plant roots. Plants use the nitrates to make nitrogen containing compounds called proteins. Some plants called **legumes**, such as peas, beans, and clover, can absorb nitrogen directly from the air in soil into their roots. Their roots have nodules containing bacteria which convert nitrogen into ammonium compounds. The bacteria 'fix' the nitrogen.

2 We eat plants such as wheat, rice, peas and beans and absorb the plant protein into our bodies. We eat animals too but they eat plants and so all our protein ultimately comes from plants.

3 Plants and animals die, animals produce manure. Bacteria in the soil convert dead and waste material back into ammonium compounds and nitrates, which can be used by new plants. Some ammonia and nitrogen escapes back into the atmosphere.

4 During flashes of lightning, nitrogen and oxygen in the atmosphere combine to make nitrogen oxide.

$$N_2(g) + O_2(g) \longrightarrow 2NO(g)$$

This is another example of the **fixation of nitrogen**. The nitrogen oxide combines with more oxygen and dissolves in rain to form very dilute nitric acid. This reacts with chemicals in the soil to form compounds called nitrates.

Intensive farming requires more nitrogen from the soil than is found naturally. Because of this, artifical fertilizers such as ammonium nitrate are added to the soil to produce very large crops. The advantage of this is that more food can be grown. The disadvantage is that some of the nitrates get washed from the soil and cause pollution in rivers and lakes.

Ammonia

Each year in Great Britain, 2.5 million tonnes of ammonia gas is made. The whole World produces 85 million tonnes. 80% of this very important chemical is made into fertilizers and the rest goes to make nitric acid, Nylon and other chemicals.

Ammonia is manufactured using the **Haber process**. (See Figure 3.) Ammonia is made by **synthesis**. That is to say, nitrogen and hydrogen are combined directly together. Both these gases are made from Natural Gas (methane). First of all, methane is heated with steam over a catalyst of nickel at a high pressure.

$$CH_4(g) + H_2O(g) \longrightarrow CO(g) + 3H_2(g)$$

methane   steam   carbon monoxide   hydrogen

Next, the carbon monoxide and hydrogen are mixed with air and again heated at high pressure over a nickel catalyst. Some of the hydrogen reacts with the oxygen in the air and leaves nitrogen.

$$O_2(g) + N_2(g) + 2H_2(g) \longrightarrow 2H_2O(g) + N_2(g)$$

from the air

Finally, the steam and carbon monoxide are removed, leaving just the nitrogen and hydrogen.

The nitrogen and hydrogen are heated to 450°C and compressed to 200 times atmospheric pressure before being passed over a catalyst of iron pellets. A proportion of the gases is converted into ammonia.

$$N_2(g) + 3H_2(g) \rightleftharpoons 2NH_3(g)$$

nitrogen   hydrogen   ammonia

The $\rightleftharpoons$ sign means that the reaction is capable of going in both directions.

It is said to be a **reversible reaction** and if left, it would reach an **equilibrium**. The temperature, pressure and catalyst help to produce as much ammonia as possible. In addition, the gases from the reaction chamber are passed over the catalyst several times and the ammonia is removed by condensing it in a refrigeration plant. This means the hot, unreacted gases are recycled.

The ammonia is transported either as a liquid at low temperature, or dissolved in water to form ammonia solution.

Figure 3
The Haber process for the manufacture of ammonia.

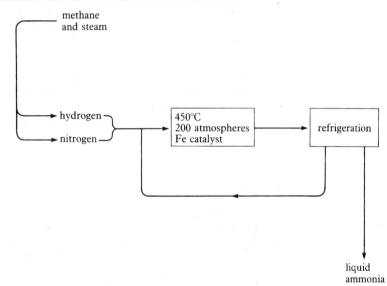

Ammonia in the laboratory

1 Ammonia is a colourless, choking gas, which is formed whenever an ammonium salt is warmed with an alkali.

For example

$$NH_4Cl(s) \;+\; NaOH(aq) \longrightarrow \; NH_3(g) \;+\; NaCl(aq) \;+\; H_2O(l)$$

ammonium   sodium   ammonia   sodium   water
chloride   hydroxide   chloride

Figure 4
What colour would you expect the litmus to turn when the contents of the tube are warmed?

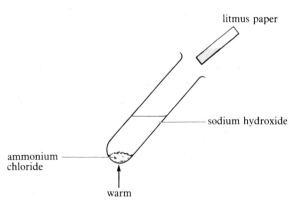

2 Ammonia is very soluble in water and forms ammonia solution, which is a weak alkali.

$$NH_3(g) + H_2O(l) \longrightarrow NH_4OH(aq)$$

Being an alkali, ammonia solution dissolves grease and is used in the manufacture of many household cleaning agents.

Figure 5
Other solvents such as petrol will dissolve grease as well as ammonia. Why wouldn't petrol be suitable for a household cleaning fluid though?

3 Ammonia reacts with acids to form salts.

For example

$$2NH_4OH(aq) + H_2SO_4(aq) \longrightarrow (NH_4)_2SO_4(aq) + 2H_2O(l)$$

4 Ammonia gas reacts with hydrogen chloride gas.

$$NH_3(g) + HCl(g) \rightleftharpoons NH_4Cl(s)$$

The $\rightleftharpoons$ sign shows that the reaction is reversible. When cooled together, ammonia gas and hydrogen chloride gas form solid ammonium chloride. When heated, ammonium chloride changes back into ammonia and hydrogen chloride. The process of a gas changing directly into a solid, missing out the liquid stage is called **sublimation**. Figure 6 shows how both reactions can happen in the same test tube.

Figure 6
Ammonium compounds sublime when heated.

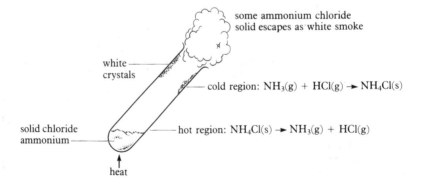

some ammonium chloride solid escapes as white smoke

white crystals

cold region: $NH_3(g) + HCl(g) \rightarrow NH_4Cl(s)$

hot region: $NH_4Cl(s) \rightarrow NH_3(g) + HCl(g)$

solid chloride ammonium

heat

Nitric acid

In the UK, three quarters of a million of tonnes of nitric acid are made each year. 30 million tonnes are made throughout the World. 90% of this is used in the production of fertilizers. The rest is used for explosives, dyes and other chemicals.

Nitric acid is made by the catalytic oxidation of ammonia. (See Figure 7.)

1 Ammonia and air are mixed and compressed to about 10 times atmospheric pressure. They are passed through a reaction chamber over a gauze made of platinum and rhodium. This gauze is a catalyst and is electrically heated to about 900°C. The ammonia is oxidised to nitrogen oxide and steam.

$$4NH_3(g) + 5O_2(g) \rightleftharpoons 4NO(g) + 6H_2O(g)$$

The reaction is reversible and the high pressure and temperature along with the catalyst make the reaction produce as much nitrogen oxide as possible.

2 The nitrogen oxide is then mixed with more air to ensure complete oxidation and passed up a tower down which water is sprayed. Nitric acid is made.

$$4NO(g) + \underset{\text{from the air}}{2O_2(g)} + 2H_2O(g) \longrightarrow 4HNO_3(aq)$$

3 60% concentrated nitric acid is produced which is then made into other chemicals such as ammonium nitrate for fertilizers and nitroglycerine for explosives.

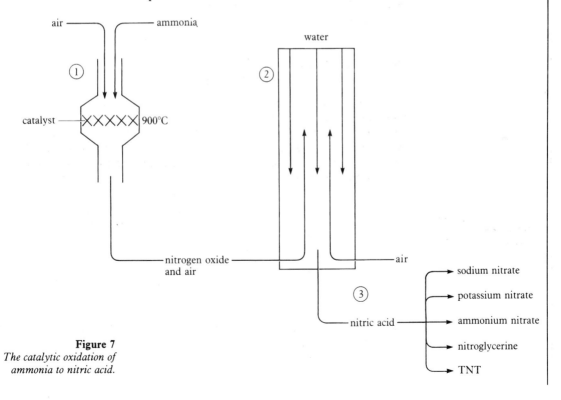

Figure 7
The catalytic oxidation of ammonia to nitric acid.

Nitric acid in the laboratory

1 Dilute nitric acid behaves like any ordinary acid:
  - it reacts with bases and alkalis to form salts
  - it reacts with carbonates to form salts and carbon dioxide.

2 With metals, nitric acid acts as an oxidising agent, producing salts and oxides of nitrogen.

For example

$$\text{zinc} + \begin{array}{c}\text{dilute} \\ \text{nitric acid}\end{array} \longrightarrow \text{zinc nitrate} + \begin{array}{c}\text{nitrogen} \\ \text{oxide}\end{array} + \text{water}$$

$$3Zn(s) + 8HNO_3(aq) \longrightarrow 3Zn(NO_3)_2(aq) + 2NO(g) + 4H_2O(l)$$

With concentrated nitric acid, nitrogen dioxide is formed instead of nitrogen oxide.

$$\text{copper} + \begin{array}{c}\text{nitric} \\ \text{acid}\end{array} \longrightarrow \begin{array}{c}\text{copper(II)} \\ \text{nitrate}\end{array} + \begin{array}{c}\text{nitrogen} \\ \text{dioxide}\end{array} + \text{water}$$

$$Cu + 4HNO_3(aq) \longrightarrow Cu(NO_3)_2(aq) + 2NO_2(g) + 2H_2O(l)$$

However, when iron or aluminium are put into concentrated nitric acid, nothing happens. This is because these metals are oxidised so rapidly by the acid, they get a protective coating of oxide which prevents further attack from the acid.

Nitrates

Most nitrates decompose when they are heated. Some give off nitrogen dioxide and oxygen and leave the oxide behind.

For example

$$\begin{array}{c}\text{Copper(II)} \\ \text{nitrate}\end{array} \longrightarrow \begin{array}{c}\text{copper(II)} \\ \text{oxide}\end{array} + \begin{array}{c}\text{nitrogen} \\ \text{dioxide}\end{array} + \text{oxygen}$$

$$\begin{array}{cccc}2Cu(NO_3)_2(s) & \longrightarrow & 2CuO(s) & + & 4NO_2(g) & + & O_2(g) \\ \text{blue crystals} & & \text{black} & & \text{brown gas}\end{array}$$

$$\text{lead(II) nitrate} \longrightarrow \text{lead(II) oxide} + \text{nitrogen dioxide} + \text{oxygen}$$

$$\begin{array}{ccccc}2Pb(NO_3)_2(s) & \longrightarrow & 2PbO(s) & + & 4NO_2(g) & + & O_2(g) \\ \text{white crystals} & & \text{yellow} & & \text{brown} & & \text{colourless} \\ & & \text{solid} & & \text{gas} & & \text{gas}\end{array}$$

When lead(II) nitrate crystals are heated, they decompose noisily. This is called **decrepitation**.

The nitrates of sodium and potassium do not decompose so easily. They have to be molten to give off oxygen (not nitrogen dioxide) and leave behind a salt called a **nitrite**.

$$\text{sodium nitrate} \longrightarrow \text{oxygen} + \text{sodium nitrite}$$

$$\begin{array}{ccc}2NaNO_3(s) & \longrightarrow & O_2(g) + & 2NaNO_2(s) \\ & & & \text{pale yellow solid}\end{array}$$

Fertilizers

All plants need certain chemicals to grow. These chemicals are found in the soil and are absorbed by the plants through their roots. Some elements are needed in tiny amounts, for example iron, manganese, boron, copper, zinc and molybdenum and are called **trace elements**. There is usually enough of these in the soil naturally and too much of them can be poisonous to the plant.

Nitrogen, potassium and phosphorus are called **major nutrients**. Plants need a lot of them for healthy growth. Without them plants may be stunted and not form roots properly. Fruiting plants may not produce their fruit if they are short of these nutrients.

The elements nitrogen, phosphorus and potassium can be added to the soil separately. Sodium or ammonium nitrate can be used to supply nitrogen. Calcium phosphate or even roasted bones are used for phosphorus. Potassium chloride is added for potassium. Often, however, all three nutrients are needed at the same time in varying amounts. A mixed fertilizer like this is called an **NPK** fertilizer.

In a fertilizer factory, ammonium nitrate is made by neutralising ammonia solution with nitric acid. Calcium phosphate or ammonium phosphate is made by reacting natural phosphate ores with sulphuric acid. Potassium chloride is mined in parts of Yorkshire or imported from Eastern Europe. Solutions of the three compounds are mixed in the required proportions and evaporated to a solid. The fertilizer is often supplied to the farmer in the form of small pellets called **prills** (Figure 8).

Figure 8   *What are the main elements in this fertilizer? Why is each one needed? Why do you think the fertilizer is made in tiny prills like this rather than in powder form?*

Key points

- The natural flow of nitrogen in the environment is called the nitrogen cycle.
- Ammonia is manufactured by the Haber process from nitrogen and hydrogen.
- Nitric acid is made by the oxidation of ammonia.
- Ammonia forms an alkaline solution with water. Ammonium compounds sublime when heated.
- Nitric acid is an oxidising agent with metals.
- Nitrates decompose when heated to give oxygen and sometimes oxides of nitrogen.
- Fertilizers are compounds containing nitrogen, potassium and phosphorus.

Quick questions

1 TNT has the formula $C_7H_5(NO_2)_3$. Why does it explode when detonated?

2 What is a catalyst? How do the iron pellets help the reaction in the Haber process?

3 What compound would be formed if ammonia gas were bubbled through a solution of potassium hydroxide?
Write an equation for the reaction.

4 Some ammonium chloride has been contaminated with sand. How could you use sublimation to purify it?

5 Explain why concentrated nitric acid can be transported in containers made of aluminium.

6 Fertilizers like ammonium and sodium nitrate often have a warning on them saying that they are a fire hazard. Why would they be dangerous if they got too hot?

7.6 *Chemicals from salt*

Salt of the Earth

200 million years ago, most of central England was covered by a big inland sea. This slowly dried out, leaving 400 000 million tonnes of salt which were gradually covered by layers of clay and sand. Today, this salt is deep under the ground and in some places, such as Winsford in Cheshire, is mined by ICI. Shafts have been sunk into the salt deposits and caverns dug out. Huge mechanical saws cut into the salt and explosives are used to free great chunks of it. Dumper trucks take 15 tonnes at a time to crushers before the salt is carried to the surface. (Figure 1.)

Figure 1
Preparing the charge for blasting rock salt in the salt mine.

More than 2 million tonnes of brown rock salt (94% sodium chloride mixed with clay) is mined each year for putting on roads in freezing weather. Elsewhere, water is pumped down into salt deposits so that brine (concentrated salt solution) is forced to the surface to be stored in great reservoirs. Liquid brine, or salt crystals evaporated from the brine are used in the chemical industry for hundreds of different products.

Salt for chlorine and sodium hydroxide

Brine is decomposed into chlorine and sodium hydroxide by the process of electrolysis in a **mercury cell**. (Figure 2.) Concentrated sodium chloride solution flows through the cell between an **anode** made of titanium and a **cathode** made of mercury which runs along the bottom of the cell. A current of up to 400 000 amps is passed between the electrodes.

At the anode, chloride ions change into chlorine gas, which is piped out of the cell and stored.

$$2Cl^-(aq) \longrightarrow Cl_2(g) + 2e^-$$

At the cathode, sodium ions change into sodium which immediately dissolves in the mercury to form an **amalgam**.

$$Na^+(aq) + 2e \longrightarrow Na \text{ (dissolved in mercury)}$$

The amalgam is carried out of the cell and mixed with water. The sodium reacts with the water to form sodium hydroxide solution. Hydrogen gas is formed as well. The mercury is returned to the cell to be used again.

$$\underset{\substack{\text{sodium dissolved} \\ \text{in mercury}}}{2Na/Hg} + 2H_2O(l) \longrightarrow 2NaOH(aq) + H_2(g) + \underset{\substack{\text{(returned} \\ \text{to cell)}}}{Hg}$$

The sodium hydroxide is either kept as a solution or evaporated to form a white solid.

Figure 2
The mercury cell. This diagram shows only a cross section of the cell. The photograph gives some idea of scale.

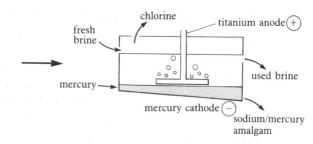

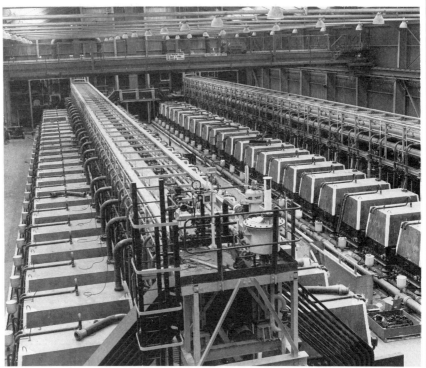

One and a half million tonnes of chlorine are produced each year in this country by this method and used in many different ways. (Figure 3.)

Figure 3
Chlorine is used to make all sorts of things. Name one product from each of the parts of this diagram.

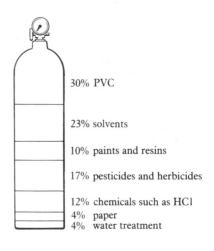

30% PVC

23% solvents

10% paints and resins

17% pesticides and herbicides

12% chemicals such as HCl
4% paper
4% water treatment

The table shows some of the compounds made from chlorine.

Compound	Use
PVC	rainwear, insulation of electric cables, drainpipes, records, plastic bottles, floor tiles, shoes, cling film
solvents such as: 1,1,1-trichloroethane dichloromethane chlorofluorocarbons	for Tipp-Ex for paint solvents for anaesthetics, refrigerants, aerosol propellants
sodium chlorate(I)	sterilising liquids like Milton and Domestos
sodium chlorate(V)	weedkiller
TCP	antiseptics
organochlorine compounds	for weedkillers and herbicides

The sodium hydroxide produced by the electrolysis of brine is an important chemical too. (Figure 4.)

Figure 4
Sodium hydroxide is added to water from industry to neutralise it before release into rivers or the sea. What sort of impurities might be in the water if an alkali is needed?

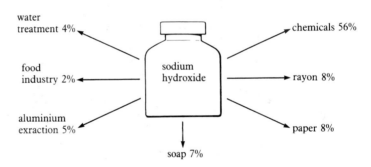

Soap

Soap is a cleaning agent. When mixed with water, it lowers the **surface tension** of the water, allowing it to spread out and 'wet' the surface of the material it is washing. At the same time, the soap molecules dislodge dirt particles and break up or '**emulsify**' grease and oil. (Figure 5.) You can find more about the structure of soap in section 9.3.

Figure 5
A soap molecule — sodium stearate.

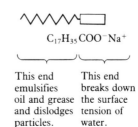

$C_{17}H_{35}COO^-Na^+$

This end emulsifies oil and grease and dislodges particles.

This end breaks down the surface tension of water.

Soap is made by heating animal fats such as beef tallow, or vegetable oils such as palm oil and coconut oil, with sodium hydroxide. The mixture is heated in a big vat with steam. The soap is made to solidify out by adding brine, and **glycerol** is an important by-product.

glyceryl stearate + sodium hydroxide ⟶ sodium stearate + glycerol
  (beef tallow)   (soap)

Bars of soap contain perfumes, colouring, deodorants and preservative as well as sodium stearate.

Chlorine in the laboratory

When concentrated hydrochloric acid is oxidised, chlorine is made (Figure 6).

$2KMnO_4$ + $16HCl(aq)$ ⟶ $2KCl(aq)$ + $2MnCl_2(aq)$ + $8H_2O(l)$ + $5Cl_2(g)$

potassium   hydrochloric   potassium   manganese(II)   chlorine
manganate(VII)   acid   chloride   chloride

The potassium manganate(VII) gives oxygen to the hydrogen atoms from the hydrochloric acid, turning them into water. The chlorine gas bubbles off as soon as the acid drops on the solid. It is collected downwards into a gas jar.

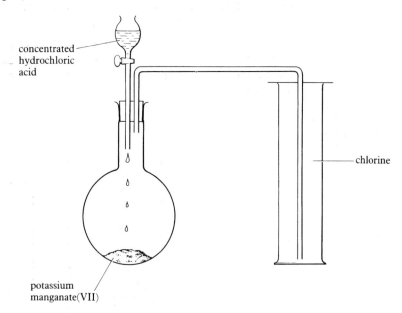

Figure 6
Making chlorine in the laboratory. Why does the delivery tube go right down to the bottom of the gas jar? What happens to the air in the gas jar?

concentrated hydrochloric acid

chlorine

potassium manganate(VII)

Chlorine is:

- heavier than air
- green in colour
- very poisonous, having a choking smell.

Chlorine is a very reactive element. In most of its reactions it changes into chloride ions as it reacts.

Chlorine reacts vigorously with metals to form compounds called chlorides. (Figure 7.)

For example

$$2Fe(s) + 3Cl_2(g) \longrightarrow 2FeCl_3(s)$$
$$\text{iron} \quad \text{chlorine} \quad \text{iron(III) chloride}$$

This is a very exothermic reaction. The iron has to be heated to start the reaction, but glows brightly once started. The iron(III) chloride is formed as a smoke which settles out to a brown solid.

Figure 7
Once the reaction has started, the Bunsen flame can be removed. Why is this?

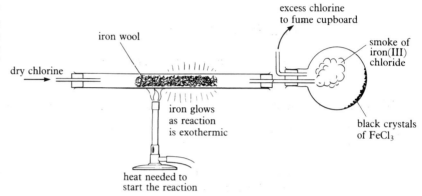

Chlorine burns in hydrogen to form hydrogen chloride gas.

$$H_2(g) + Cl_2(g) \longrightarrow 2HCl(g)$$

Hydrogen chloride gas is very soluble in water and as soon as it comes into contact with damp air, it changes from an invisible gas to a mist of hydrochloric acid droplets. Hydrochloric acid is manufactured in this way. (Figure 8.)

Figure 8
The manufacture of hydrochloric acid.

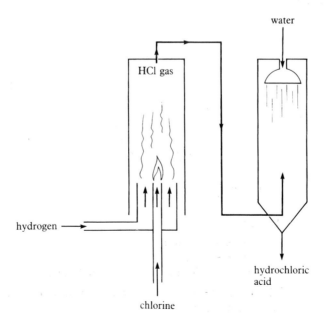

Chlorine dissolves in water to form a bleach. (Figure 9.)

$$Cl_2(g) + H_2O(l) \longrightarrow HCl(aq) + \underset{\text{chloric(I) acid}}{HOCl(aq)}$$

Chlorine is also added to domestic water supplies too, to kill harmful germs. In swimming pools, chlorine, or sodium chlorate(I) is used to keep the water clean, although in more modern pools, ozone is used instead.

Figure 9
*A test for chlorine. Do
any other gases change
litmus in this way?*

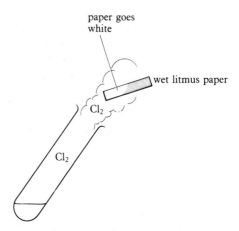

paper goes
white

wet litmus paper

Cl_2

Cl_2

Chloric(I) acid is an unstable acid. It reacts with vegetable dyes to bleach them and germs to kill them. It bleaches and kills germs by oxidation.

$$\underset{\text{chloric(I) acid}}{HOCl(aq)} \longrightarrow HCl(aq) + O$$

this oxygen
oxidises the colour
or germs

If chlorine is dissolved in sodium hydroxide solution, salts of the two acids are formed.

$$Cl_2(g) + 2NaOH(aq) \longrightarrow NaCl(aq) + \underset{\text{sodium chlorate(I)}}{NaOCl(aq)}$$

Sodium chlorate(I) is an oxidising agent as well. It is often known by the common name of **sodium hypochlorite** and is used in Domestos, and in very dilute solution, in liquids used to sterilise babies' bottles.

Group VII of the periodic table — the halogens

The members of Group VII of the periodic table are very similar in their properties. Each of them has seven electrons in the outer shell, and when they react, they each take in one extra electron to form a negative halide ion.

$$\left.\begin{array}{ll}\text{fluorine} & F_2 \\ \text{chlorine} & Cl_2 \\ \text{bromine} & Br_2 \\ \text{iodine} & I_2\end{array}\right\} X_2 + 2e \longrightarrow 2X^- \left\{\begin{array}{ll}\text{fluoride ion} & F^- \\ \text{chloride ion} & Cl^- \\ \text{bromide ion} & Br^- \\ \text{iodide ion} & I^-\end{array}\right.$$

Here are some properties that the halogens have in common.

1 They are all coloured:

> fluorine is a yellow gas
> chlorine is a green gas
> bromine is a red liquid with a brown vapour
> iodine is a black solid with a violet vapour.

2 They all dissolve in water to form a solution which bleaches:

> fluorine reacts violently
> chlorine and bromine dissolve well
> iodine is the least soluble in water but dissolves well in ethanol to make an antiseptic.

3 They all react with metals such as iron wool:

> iron wool bursts into flames in fluorine
> iron reacts exothermically with chlorine and bromine vapour
> iron reacts very slowly with iodine vapour.

Sodium carbonate

Sodium carbonate is manufactured from salt and limestone. It is used to make glass and chemicals for the food and drink industry and also in textile and dye manufacture. **Sodium hydrogencarbonate** is also made in the same process and is used in **baking powder**.

When sodium hydrogencarbonate is heated, it decomposes into sodium carbonate and carbon dioxide.

$$2NaHCO_3(s) \longrightarrow Na_2CO_3(s) + CO_2(g) + H_2O(l)$$
$$\text{sodium} \qquad\qquad \text{sodium}$$
$$\text{hydrogencarbonate} \quad \text{carbonate}$$

If baking powder is put into a cake mixture, the carbon dioxide puffs the cake up whilst it is being cooked, making the cake light instead of solid and heavy.

Glass

Glass is made by melting together the following chemicals:

> sodium carbonate ⎫
> limestone ⎪ at a temperature of 1500°C
> sand ⎬ in a furnace heated by gas
> recycled glass ⎪ or electricity
> (cullet) ⎭

These substances react to form a mixture of **sodium silicate** and **calcium silicate**. Other chemicals are added: **manganese(IV) oxide** makes it clear, cobalt oxide makes blue glass and boron oxide makes very tough glass called borosiliate glass (Pyrex).

The molten glass that oozes out of the furnace is cut and moulded in machines worked by compressed air. (Figure 10.) The newly formed bottles and jars have to be cooled down very slowly over a period of many hours so that stresses are not set up within the glass. Stresses would weaken the glass and it would break easily.

Figure 10
The very hot glass is blown into shape by compressed air. Newly formed bottles must be cooled down very slowly so that the glass does not crack.

Figure 11
This huge sheet of plate glass for shop windows was made by floating molten glass on molten tin to form a flat surface. This is called float glass. Why does the glass float on the tin?

Key points

- Salt is mined as rock salt or pumped out from under the ground as brine.
- The electrolysis of brine makes chlorine and sodium hydroxide.
- Many products are made from chlorine and sodium hydroxide.
- Soap is made when oils and fats are boiled with sodium hydroxide.
- Members of Group VII are called the halogens. They have similar properties.
- Brine is used in the manufacture of sodium carbonate which in turn is used to make baking powder and glass.

Quick questions

⋆ 1 Name the materials used for the anode, cathode and electrolyte in the manufacture of sodium hydroxide by electrolysis.

2 Suggest one use for the hydrogen made as a by-product in the manufacture of sodium hydroxide.

3 Sodium hydroxide is a deliquescent solid. What does deliquescent mean?

4 Name three elements found in anaesthetics and aerosol propellants.

5 What do you think the starting materials were for the manufacture of 'Palmolive' soap?

6 Suggest a possible reason why tap water smells more strongly of chlorine in the summer.

7.7 Sulphur and its compounds

Rubber

The connection between sap from a rubber tree and a modern car tyre, is sulphur. **Latex** is the milky liquid which seeps out from under the bark of the rubber tree when it is cut. (Figure 1.) It solidifies to form a soft, pliable, but not very bouncy rubber. Latex is made of chains of carbon atoms all twisted and tangled up. Sulphur atoms are put between the chains to make links rather like bed springs. This process is called **vulcanisation**. When the vulcanised rubber is stretched, the long chains are stretched out, but are pulled back into shape by the sulphur links. It's sulphur that adds the bounce!

Sulphur is not the only substance added to the rubber in tyre manufacture. **Pigments** are added to colour it. Carbon is added as a **reinforcing agent** to make it more resistant to wear. **Antioxidants** are added to stop air and other gases, such as small amounts of ozone, from attacking the sulphur links. The rubber and its additives are put into a big hydraulic press, heated by steam, and squashed to a force of nearly one tonne per square centimetre. (Figure 2.)

Figure 1 *The sap from this rubber tree is the latex from which rubber is made.*

Figure 2 *Very high pressure is needed to make sulphur react with latex to vulcanise rubber.*

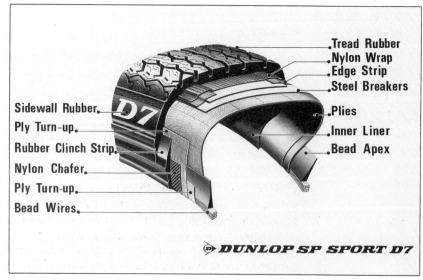

Figure 3
The rubber you see on the outside of a tyre is only part of the tyre's structure. Steel is used to make the rim and fabrics such as nylon and cotton are used to strengthen the body of the tyre.

Sidewall Rubber

Ply Turn-up

Rubber Clinch Strip

Nylon Chafer

Ply Turn-up

Bead Wires

Tread Rubber

Nylon Wrap

Edge Strip

Steel Breakers

Plies

Inner Liner

Bead Apex

D7

⊕ *DUNLOP SP SPORT D7*

A native element

Unlike most metals and other non-metals such as phosphorus and chlorine, sulphur is found in the ground in its **native**, or natural state. Most of the World's sulphur comes from underground deposits in Texas, Mexico, Sicily, the USSR and Poland. Sulphur is also extracted from crude oil as an impurity.

Because sulphur has a fairly low melting point (about 114°C) it can be extracted from the ground with hot water using a **Frasch pump**. (Figure 4.) A hole is drilled down into the sulphur bed (usually about 150 metres down) and the pump, consisting of three concentric pipes is put in. Water under pressure at 170°C is pumped down the outer tube to melt the sulphur. Compressed air is forced down the central tube to force the water and molten sulphur up through the central tube.

Figure 4
The Frasch sulphur pump. Why do you think the sulphur is kept in the inner-most tube and the hot water is kept on the outside?

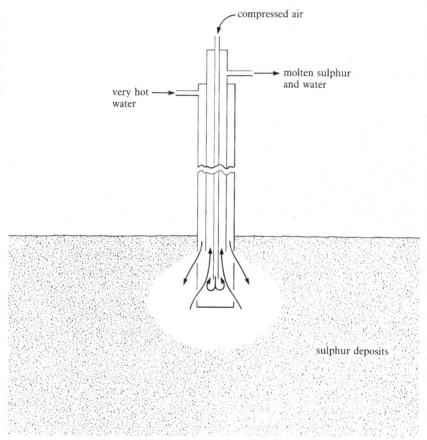

compressed air

molten sulphur and water

very hot water

sulphur deposits

Like carbon, sulphur has **allotropes**. It can exist as two different crystalline forms, each having a different shape. Below 96°C, small diamond shaped crystals called **rhombic sulphur** are formed. Above this temperature, long needle-like crystals called **monoclinic sulphur** are formed. The temperature of 96°C at which rhombic crystals change into monoclinic crystals is called the **transition temperature.** (Figure 5.)

Figure 5 *These allotropes of sulphur are different because the sulphur atoms are arranged in different ways.*

Sulphur dioxide

When sulphur is heated, it melts and burns with a dark blue flame to form sulphur dioxide gas.

$$S(s) + O_2(g) \longrightarrow SO_2(g)$$
$$\text{sulphur dioxide}$$

This gas is also made when salts called **sulphites** are added to a dilute acid. For example:

$$Na_2SO_3(s) + 2HCl(aq) \longrightarrow 2NaCl(aq) + SO_2(g) + H_2O(l)$$

sodium sulphite + hydrochloric acid ⟶ sodium chloride + sulphur dioxide + water

Figure 6 shows the apparatus that would be used.

Figure 6
Sulphur dioxide is heavier than air and so is collected downwards into the test tube. Why isn't the gas bubbled through a beaker of water and collected like other gases?

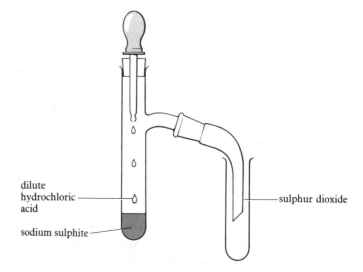

dilute hydrochloric acid

sodium sulphite

sulphur dioxide

Sulphur is a colourless, choking gas with a metallic taste. It can be liquefied when compressed to about three times atmospheric pressure, and so can be kept in a metal canisters, just as butane (camping gas) can. (Figure 7.)

Figure 7
Sulphur dioxide can be liquefied by pressure.
• *Why is it usually transported in this way?*

Since it is the oxide of a non-metal, sulphur dioxide dissolves in water to form a solution of an acid called **sulphurous acid**.

$$SO_2(g) + H_2O(l) \longrightarrow H_2SO_3(aq)$$
sulphurous acid

This is what makes you cough if you breath the gas by accident.

Uses and problems

Sulphur dioxide is a useful gas. When added to wood pulp, it bleaches it white and so sulphur dioxide and sodium sulphite are very important in the paper industry.

When added to foods such as fruit, sulphur dioxide makes the food keep longer by killing bacteria. Sulphur dioxide is a preservative.

Bleaching wood pulp and preserving fruit with sulphur dioxide both involve reduction reactions. The sulphurous acid formed from the sulphur dioxide and water is an unstable compound and readily takes in extra oxygen to form sulphuric acid.

$$H_2SO_3(aq) + O \longrightarrow H_2SO_4(aq)$$
sulphurous acid   oxygen   sulphuric acid

Removal of oxygen from wood pulp makes it white. Lack of oxygen in the fruit eventually causes the bacteria to die.

Acid rain

The effects of **acid rain** are easy to see. Stone buildings become eroded and trees die. Lakes in Scandinavia have become so acidic that all the fish die and water can no longer support life.

Figure 8
*The effects of acid rain
are easy to see.*

The cause of acid rain is much more difficult to understand. First of all, power stations that burn coal or oil emit large quantities of sulphur dioxide into the atmosphere because these fossil fuels contain sulphur and sulphur compounds. Up to two thirds of this gas may become dissolved in the rain and fall as dilute sulphuric acid.

$$SO_2(g) + H_2O(l) \longrightarrow \quad H_2SO_3(aq)$$
$$\text{sulphurous acid}$$

and then

$$H_2SO_3(aq) + \text{oxygen} \longrightarrow H_2SO_4(aq)$$

Levels of acidity are measured on the **pH scale**. Neutral water has a pH of 7. Normal rain water has a value of 5 (because of naturally dissolved carbon dioxide). Acid rain has pH values from 5 to 2.5 in some of the worst cases. Ordinary laboratory dilute sulphuric acid has a pH of 1, so some acid rain is really quite corrosive. Some estimates suggest that 5 million tonnes of sulphur dioxide are blown over the North Sea to Scandinavia and yet this is only about 40% of the total sulphur dioxide produced in Europe as a whole.

Power stations emit a lot of sulphur dioxide, but that is not the only cause of acid rain. The intense heat inside power station furnaces makes oxygen and the normally unreactive nitrogen combine to make **oxides of nitrogen**. Like sulphur dioxide, they dissolve in rain to make dilute nitric acid, increasing the acidity. Cars and lorries make oxides of nitrogen too and emit it in their exhaust fumes. These, along with **unburnt hydrocarbons** from the petrol produce ozone and organic chemicals in the upper atmosphere, when sunlight falls on them. This mixture of chemicals and chemical reactions produces acid rain.

The cure for acid rain is to stop producing sulphur dioxide and oxides of nitrogen. EEC countries have agreed among themselves to reduce sulphur dioxide output by 60% by the year 1995, but it will be an expensive process. The sulphur dioxide can be washed out of the fumes from power station chimneys before they reach the atmosphere. New furnances can be designed that burn pulverised coal at lower temperatures so that fewer oxides of nitrogen are formed. Petrol for cars could be made lead free and catalyst converters could be fitted to exhaust systems to remove oxides of nitrogen and unburnt hydrocarbons. Will it all happen before you read this book, I wonder?

Sulphuric acid

Each year, 130 million tonnes (100 000 swimming pools-ful!) of sulphuric acid are made throughout the World. Look at Figure 9.

Figure 9
Sulphuric acid is a very important chemical.

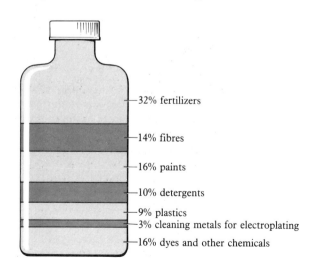

- 32% fertilizers
- 14% fibres
- 16% paints
- 10% detergents
- 9% plastics
- 3% cleaning metals for electroplating
- 16% dyes and other chemicals

In Great Britain, sulphuric acid is manufactured by the **Contact process**. (Figure 10.) Sulphur dioxide from burning sulphur is mixed with air. These gases are very slightly pressurised and heated to a temperature of 420°C before being passed over a catalyst of **vanadium(V) oxide**. They react to form sulphur trioxide.

$$2SO_2 \quad + O_2 \rightleftharpoons \quad 2SO_3(g)$$

sulphur dioxide   sulphur trioxide

The catalyst and temperature ensure that 99% of the sulphur dioxide and oxygen react.

The sulphur trioxide is dissolved in concentrated sulphuric acid to form a thick oily liquid called **oleum**. (Dissolving it directly in water would create a lot of heat and cause a mist of sulphuric acid droplets in the air.) The oleum is then carefully diluted with water.

$$H_2SO_4 . SO_3(l) + H_2O(l) \longrightarrow 2H_2SO_4(aq)$$

oleum

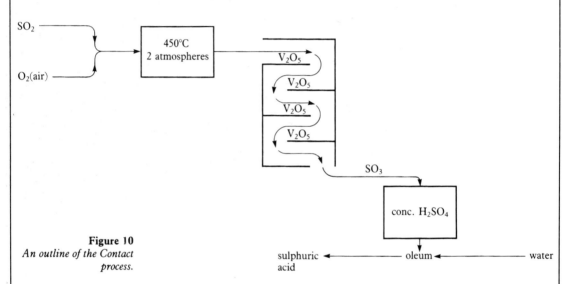

Figure 10
An outline of the Contact process.

Sulphuric acid in the laboratory

Dilute sulphuric acid is an ordinary acid. It reacts with

- metals to form a salt and hydrogen
- bases to form a salt and water
- alkalis to form a salt and water
- carbonates to form a salt, carbon dioxide and water.

In each case, the salt formed is a **sulphate**.

Concentrated sulphuric acid is an oily liquid, much heavier than water and despite its name, it doesn't have the same chemical reactions as an ordinary acid. This is because it contains covalent molecules and no hydrogen ions. For example, when litmus paper is dipped into concentrated sulphuric acid, it chars and falls to pieces. When a piece of magnesium ribbon is put into the liquid, no reaction takes place.

Concentrated sulphuric acid reacts very vigorously with water. If concentrated sulphuric acid is poured quickly into water, the heat given out as the covalent molecules are changed into ions, makes the water boil.

When diluting concentrated sulphuric acid, always wear gloves and a face mask. Always add the acid to the water, slowly. Never add water to acid.

Concentrated sulphuric acid absorbs water so readily that it is a good drying agent for gases that contain water vapour. (Figure 11.)

Figure 11
When a gas is passed through this apparatus the sulphuric acid at the bottom of the flask removes any water vapour.

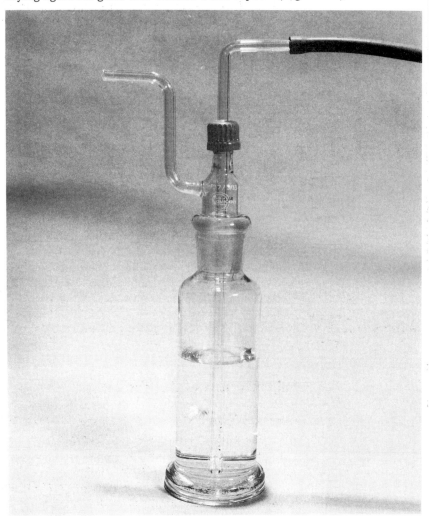

Concentrated sulphuric acid will also remove the water from sugar to leave just carbon. It is a **dehydrating agent**. Sugar (sucrose) is a **carbohydrate**, and when water is removed, only carbon is left. (Figure 12.)

$$C_{12}H_{22}O_{11} \longrightarrow 12C + 11H_2O$$
(removed by the acid).

Figure 12 *Concentrated sulphuric acid dehydrates sugar and leaves only carbon. The reaction is exothermic.*

Key points

- The process of adding sulphur to latex to put the bounce in it is called vulcanisation.
- Sulphur is extracted from underground deposits with a Frasch pump.
- Sulphur exists in different crystalline forms called allotropes; rhombic and monoclinic sulphur.
- Sulphur dioxide is used as a bleach and a food preservative.
- Sulphuric acid is manufactured by the Contact process.
- Sulphuric acid has numerous uses. When concentrated it is a dehydrating agent.
- Acid rain is caused by sulphur dioxide and oxides of nitrogen.

Quick questions

1 Write the word equation for the reaction between dilute nitric acid and potassium sulphite.

2 (a) Explain how sulphur dioxide bleaches wood pulp.
(b) Newspaper left in the sun (and in the air) goes brown.
Explain the chemistry of what is happening.

3 What is a catalyst. Why doesn't the vanadium(V) oxide in the Contact process need replacing?

4 Copper sulphate crystals contain water of crystallisation. Explain what you think will happen if concentrated sulphuric acid is added to some of them. What would you see?

5 Explain and describe what would happen if dilute sulphuric acid is added to (a) zinc (b) copper oxide (c) magnesium carbonate.

7.8 *Water*

Water supply

Our water supply is part of the **water cycle**. Rain water, originally evaporated from the sea and lakes by the sun's heat, falls as rain on hills. The rain seeps into streams and lakes and some of it soaks into the ground where it may be recovered from wells and boreholes. Most of it runs back into the sea, and the water cycle starts again.

Figure 1
The water cycle.

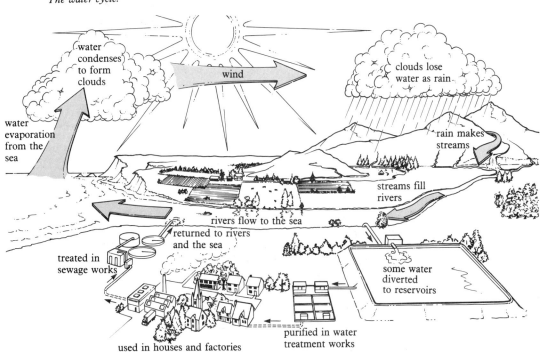

We take the cleanliness of our water supply for granted. However, before it reaches the taps in our homes, the water will have been in contact with the ground as it flows in streams and rivers, and will contain many dissolved substances and also sediments of clay and grit. It may even contain harmful bacteria.

Water arrives at the water treatment plant from reservoirs, rivers or bore holes and it needs to be purified. First of all it is passed through **screens** to remove large pieces of debris. Next, it is allowed to flow slowly through **settling tanks** so that heavy sediment and small pieces of grit sink out of the water. A chemical called alum is added at this stage to make small particles of silt and clay stick together and come out of suspension. This process is called **coagulation**. The water then flows through more settling tanks and then through **filter beds** of sand to remove as much undissolved solid as possible.

The water will probably be aerated to remove unpleasant smells. Finally, the water has a small amount of **chlorine** gas added to kill any harmful bacteria and then the water will be safe to drink. It is then stored in covered **reservoirs** until it is needed.

Sewage treatment

Water that has been used in houses, hospitals and industries must be treated at the **sewage works** before it can be returned to the water cycle.

At the sewage works, large solid debris is removed by **screens**, just as at the water treatment plant. In **settling tanks,** the sewage is slowly stirred so that much of the solid waste slowly sinks to the bottom as a thick sludge. This sludge is then put into big storage tanks where bacteria break it down into a harmless solid which can be put onto the land as fertilizer, or dumped at sea. Whilst the sludge is decomposing, **methane** gas is made, which is sometimes used as a fuel at the sewage works.

The water from the settling tanks is then trickled through **filter beds** of clinker. Air is absorbed by the water during this process so that micro-organisms that live on the clinker can break down any organic matter still in the water.

Finally, the water is passed through several sand filters before being put back into the river. Whilst not pure enough to drink, it is no longer a health danger.

Figure 2
A sewage works. The circular parts are filter beds. Can you see where the treated water goes back into the river?

Hardness of water

When soap is shaken with pure water, a frothy lather is formed. Tap water is pure enough to drink, but it is not pure in a chemical sense. It contains dissolved chemicals. Some of these are a nuisance because they react with soap during washing and turn it into an insoluble substance called **scum**. Not only is scum unsightly in the wash basin, it also uses up extra soap that could have been used for washing. Water that does not lather easily with soap and which forms a scum is said to be **hard**.

Rain water is slightly acidic because it dissolves carbon dioxide as it falls through the air. If this acidic solution falls on limestone rocks, the limestone dissolves to form a dilute solution of a salt, **calcium hydrogencarbonate**.

$$CaCO_3(s) \ + \ CO_2/H_2O(aq) \longrightarrow Ca(HCO_3)_2(aq)$$

limestone   rain water   calcium hydrogencarbonate solution

When soap is added to this solution the calcium ions combine with the soap ions and form an insoluble compound.

sodium stearate(aq)	+	Ca^{2+}(aq) calcium ions	$\longrightarrow$	calcium stearate(s)	+	Na^+ sodium ions
soap		in hard water		scum		from the soap

Water containing calcium hydrogencarbonate is said to have **temporary hardness** because when heated, the solution decomposes and forms solid calcium carbonate. This takes the calcium ions out of solution and makes the water soft again.

$$Ca(HCO_3)_2(aq) \xrightarrow{\text{heat}} CaCO_3(s) \ + \ CO_2(g) \ + \ H_2O(l)$$

If this decomposition happens inside central heating pipes, or in a kettle, then calcium carbonate is deposited on the sides of the pipes or kettle. It is then called **scale** or fur, and it is a nuisance because it can block the pipes or cut down the efficiency of the heating element in an electric kettle. If hard water of this sort is used for washing and cleaning in industry, this process can block water jets or prevent electroplating sticking to metal surfaces.

Figure 3 *This chemical dissolves scale — calcium carbonate. What sort of compound must the chemical be?*

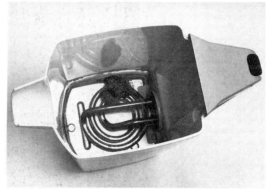

Figure 4 *This device stops scale forming in an un-chemical way. How does it work?*

If rain water dissolves calcium sulphate from rocks, the hardness formed is called **permanent hardness** because this compound does not decompose when heated. Water often contains temporary and permanent hardness.

Caves

In limestone areas, rivers that have flowed for hundreds of thousands of years have slowly cut into the rock, forming underground streams and sometimes huge caverns. Where water containing calcium hydrogencarbonate has dripped from the roofs of these caves, small deposits of calcium carbonate have built up. Where the drips hit the ground and evaporate, small pillars of calcium carbonate appear too. Upwards growths like these are called **stalagmites** and downward growths are called **stalactites**. Their rate of growth is very slow, just a few millimetres in a thousand years.

Figure 5
These beautiful formations have been made from hard water.

Removing hardness of water

Scale can be removed from kettles by dissolving it away with a weak acid, but it is better not to let it form in the first place. This is achieved by removing the calcium ions from the water before it is used. One way is to add a water softener such as sodium carbonate (**washing soda**). This combines with the calcium ions in the water turning them into insoluble calcium carbonates and removing them from solution.

$$Na_2CO_3(aq) + Ca^{2+}(aq) \longrightarrow CaCO_3(s) + 2Na^+(aq)$$

| washing soda | hard water | calcium carbonate | soft water |

Alternatively, the hard water can be passed through a **de-ionising resin**. This consists of tiny beads of a chemical that removes unwanted calcium ions and swaps them for sodium ions. Completely pure water, called **de-ionised water**, can be made in this way too. The resin removes all ions and replaces them with hydrogen ions and hydroxide ions. Washing powders contain compounds called **polyphosphates** which remove unwanted calcium ions from hard water in a similar sort of way, making the water soft for washing machines and dishwashers.

Figure 6
This chemical contains polyphosphates which soften water. Polyphosphates are a problem in rivers and lakes if used in excess.

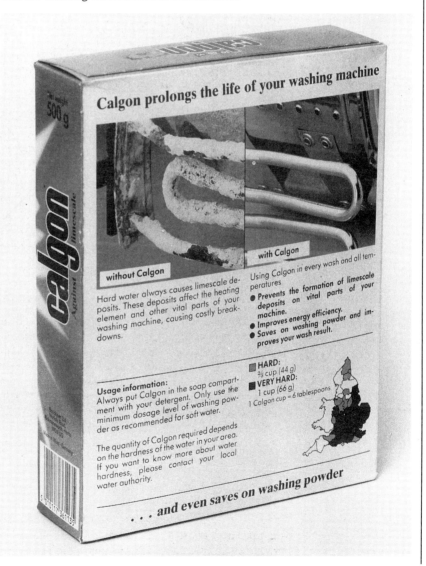

Nitrates and phosphates

Water draining into rivers and lakes from farmland often contains large amounts of **nitrates**. These come from fertilizers washed from the soil by rain. High levels of nitrates in drinking water can be a health hazard. They are particularly dangerous for young babies because they reduce the oxygen level in their blood.

Water from sewage works often contains **phosphates** from detergents. Nitrates and phosphates in rivers and lakes cause problems. Being fertilizers, they encourage the growth of green plants, and in particular, **green algae**, which grow very quickly. When these plants die, the bacteria that decompose them take large amounts of oxygen from the water. Eventually the water becomes so short of oxygen that fish and water life cannot survive and the river or lake dies. This process is called **eutrophication**.

Figure 7
Phosphates and nitrates have caused the algae in this lake to grow very fast. Not only has this blocked out light from other plants, but it has caused the lake to become eutrophicated.

Key points

- Water continuously circulates in a process called the water cycle.
- Water is made fit for drinking at water treatment plants. Sewage is made clean enough to put back into rivers or the sea at sewage works.
- Hardness of water is caused by calcium ions. These turn soap into scum.
- Water may be softened by using washing soda, polyphosphates or de-ionising resins.
- Temporary hard water is caused by calcium hydrogencarbonate solution. This decomposes into scale when heated.
- Water with large amounts of nitrates and phosphates in it can become eutrophicated.

Quick questions

1 Tap water is safe to drink, but not pure. Explain what is meant by this. Describe an experiment that you could do to demonstrate it.

★ 2 When a sample of water was shaken with 40 cm^3 of soap solution, it did not lather but formed a grey scum. Another sample of the same water was boiled and cooled and then shaken with soap. A frothy lather was formed after only 20 cm^3 of soap had been added. A third sample of distilled water lathered with only 1 cm^3 of soap. Explain these results.

7.9 Analysis of non-metal compounds

Chemists sometimes have to be detectives. They need to know just what is in a substance. They have to perform analysis on it. The analysis might be as simple as finding out the amount of salt in a food sample, or as complicated as detecting minute amounts of drugs on someone's clothing. This chapter is about analysis that uses some of the chemicals that you have met in this Section. Use it as a key.

Gases

1 (a) the gas is coloured .... go to 2
  (b) the gas is colourless .... go to 3

2 (a) the gas is brown .... nitrogen dioxide, NO_2
  (b) the gas is green, has a choking smell
      and bleaches wet litmus paper .... chlorine, Cl_2

3 (a) the gas has a smell .... go to 4
  (b) the gas has no smell .... go to 5

4 (a) the gas has a metallic taste, makes you sneeze and
      turns paper soaked in potassium dichromate(VI)
      solution from yellow to green .... sulphur dioxide, SO_2
  (b) the gas has a choking smell and turns
      damp litmus paper blue .... ammonia, NH_3
  (c) the gas is misty, has a choking smell
      and turns damp litmus paper red .... hydrogen chloride, HCl

5 (a) the gas puts out a lighted splint and turns
      lime water cloudy .... carbon dioxide, CO_2
  (b) the gas relights a glowing splint .... oxygen, O_2
  (c) the gas explodes when lit .... hydrogen, H_2

Solids that react with acids and alkalis

1 Add dilute hydrochloric acid to the solid and identify the gas given off.
  (a) Carbon dioxide evolved .... carbonate, CO_3^{2-}
  (b) Sulphur dioxide evolved .... sulphite, SO_3^{2-}

2 Add sodium hydroxide solution and warm gently.

(a) Ammonia gas evolved .... ammonium compound, NH_4^+

3 Add a few drops of concentrated sulphuric acid and warm very gently.

(a) Hydrogen chloride gas evolved .... chloride, Cl^-

(b) light brown, acidic fumes of nitric acid vapour .... nitrate, NO_3^-

Compounds identified by their ions in solution

Make a solution of the compound in distilled water.

1 Add a few drops of silver nitrate solution.

(a) a white precipitate, soluble in ammonia solution .... chloride, Cl^-

(b) cream precipitate, only soluble in concentrated
ammonia solution .... bromide, Br^-

(c) yellow/green precipitate, insoluble
in ammonia solution .... iodide, I^-

2 Add a few drops of barium chloride solution.

(a) white precipitate, insoluble in
dilute hydrochloric acid .... sulphate, SO_4^{2-}

Key points

- Gases can be identified by their smell, colour and effect on litmus.
- Some compounds can be identified by the gases they give off when reacted with acids or alkalis.
- Other compounds can be identified when specific reagents are added to their solutions.

Quick questions

⋆ **1** Identify these gases without looking back into the book,
(a) no smell, no colour, pops in a flame
(b) no smell, no colour, turns limewater cloudy
(c) makes you cough and sneeze, turns potassium dichromate(VI) green
(d) the only alkaline gas.

⋆ **2** A white powder fizzes when dilute hydrochloric acid is added to it. The gas evolved puts out a flame. When the powder is added to some sodium hydroxide solution and warmed, a colourless gas is given off which makes you cough and turns litmus paper blue. Name the white powder.

⋆ **3** Some colourless crystals are dissolved in water. If barium chloride solution is added to the solution a white precipitate is formed. If the crystals are added to dilute hydrochloric acid, they fizz and give off a pungent gas that makes you sneeze. Further analysis shows that the crystals contain calcium ions. Name the compound.

★ 4 When concentrated sulphuric acid is added to a solid, misty fumes are evolved which are acidic to litmus. When some of the gas is bubbled through silver nitrate solution, a white precipitate is formed which quickly dissolves in ammonia solution. Identify the ion in the compound and explain everything that has been going on.

Questions

1 The following table provides some information about the gases in the air.

Substance	Melting point (°C)	Boiling point (°C)	% by volume in the air
argon	−189	−186	0.93
carbon dioxide	sublimes at −78°C		0.03
helium	−270	−269	0.0005
krypton	−157	−152	0.0001
neon	−249	−246	0.0015
nitrogen	−210	−196	78.03
oxygen	−219	−183	20.99
xenon	−112	−108	0.000008

The melting points and boiling points were measured at atmospheric pressure.)

(a) Answer the following questions. Each substance may be used once, more than once or not at all.

From the substances shown, name

(i) the substance which occurs least in the air,
(ii) the substance which is a liquid over the greatest range of temperature,
(iii) **one** substance which is a liquid at −200°C,
(iv) **one** substance which is a gas at −200°C,
(v) the substance most likely to increase in amount in a crowded room.

NEA

★ 2 Both egg shell and oyster shell contain calcium carbonate. This question is about an investigation into whether they both contain similar amounts of calcium carbonate.

The calcium carbonate in the shells can be measured by reacting it with acid.

$$CaCO_3(s) + 2HCl(aq) \longrightarrow CaCl_2(aq) + CO_2(g) + H_2O(l)$$

(a) Why would a mixture of calcium carbonate and acid lose mass as it reacted?
(b) When acid is added to calcium carbonate a fine spray is often given off. Why would this cause an error in this investigation?

(c) Draw a labelled diagram of the apparatus you would use to carry out the reaction between acid and egg shell or oyster shell. List carefully the measurements you would make to carry out the investigation.

(d) How would you make sure that
  (i) the reaction between the acid and the shell has finished, and
  (ii) that there is no more calcium carbonate left in the shell?

(e) How would your results tell you about the amounts of calcium carbonate in egg shell and oyster shell?

SEG

3 The diagram shows apparatus which can be used to find the composition of the air. 100 cm³ of air were placed in syringe A with syringe B empty. The copper was heated strongly and the air was passed to and from syringes A and B over the hot copper and finally returned to syringe A.

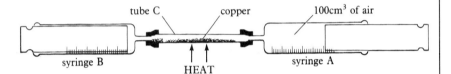

(a) (i) Which gas does copper remove from the air?
  (ii) Name the compound that is formed.
  (iii) Describe the colour change you would observe during the reaction.
  (iv) Which would be the most abundant gas in the mixture remaining in syringe A at the end of the experiment?

(b) The gas involved in the reaction is also used up during breathing.
  (i) Describe in chemical terms another process which would result in this gas being removed from the air.
  (ii) What volume of gas would be present in a 100 cm³ sample of air?
  (iii) Describe how you would test for the presence of this gas and state the result you would expect.

LEAG

4 When ammonium nitrate is used as a fertilizer only about half is used by the crop:

(a) Use the table below to draw a pie diagram of what happens to this fertilizer.

10%	washed out by rain
10%	decomposed
25%	remain in soil
55%	used by crops

(b) (i) Why is ammonium nitrate sometimes found in river water?
  (ii) Give two reasons why ammonium nitrate in river water is a pollutant.

★   **(c)** (i) Work out the formula mass of ammonium nitrate (NH_4NO_3).
            (Relative atomic masses are N 14, H 1, O 16.)
        (ii) Work out the percentage of nitrogen in ammonium nitrate.

SEG

★ **5 (a)** Catalysts are used in many industrial processes, for example, in the
        production of ammonia (NH_3).

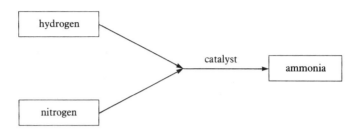

        (i) Name the transition element that is used as the catalyst.
        (ii) Give two reasons why catalysts are important for many industrial
            processes.
        (iii) Write a balanced equation for the production of ammonia.

   (b) The ammonia formed is in a mixture with unreacted nitrogen and
        hydrogen.

Gas	Boiling point/°C
ammonia	−34
nitrogen	−196
hydrogen	−253

   (i) Use the boiling points of these gases to explain how ammonia can be
        separated from the unreacted nitrogen and hydrogen.
   (ii) After ammonia is separated what use is made of the unreacted
        nitrogen and hydrogen?

   (c) A large amount of ammonia is converted into nitric acid (HNO_3).

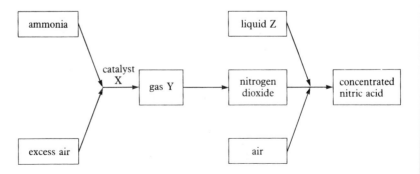

        Give the name of
        (i) catalyst **X**;
        (ii) gas **Y**;
        (iii) liquid **Z**.

(d) Much of the nitric acid produced is reacted with ammonia to form the fertilizer known as 'Nitram'. Give the chemical formula of 'Nitram'.

(e) 'Nitram' is sometimes coated with calcium carbonate and sold as 'Nitro-chalk'. 'Nitro-chalk' acts as a fertilizer and also neutralises acid soils.
  (i) Why are fertilizers added to soil?
  (ii) Why is there an increased world demand for fertilizers each year?
  (iii) Explain how 'Nitro-chalk' neutralises acid soils.

SEG

6 A company called **Brinetech** makes chemicals from rock salt (sodium chloride) which is found in rocks underneath Cheshire. This is a sketch of the factory.

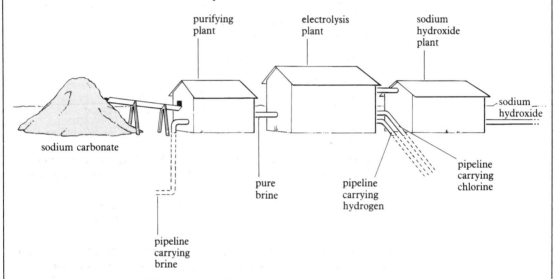

The sodium chloride arrives at the factory by pipeline as brine.

(a) What is brine?

(b) What would be the **simplest** way of obtaining brine from the underground rock salt deposits?

(c) (i) Where in Britain would you expect Brinetech to have built their factory?
  (ii) Explain why it would be sensible to build the factory there.

(d) The brine contains calcium salts as impurities. These are removed by adding solid sodium carbonate to the brine.
  (i) Complete this word equation to explain how the solid sodium carbonate reacts with the impurities.

    sodium carbonate + calcium chloride ⟶ sodium chloride + ...

  (ii) After this reaction the solid unwanted product must be removed from the brine. What method would you suggest to do this?

(e) In the electrolysis plant there are many electrolysis cells like this in which the sodium chloride is electrolysed.

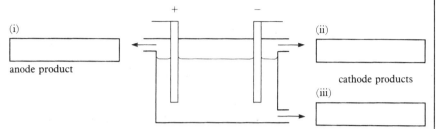

(i) anode product

(ii) cathode products

(iii)

Name the three products in the boxes on the diagram.

(f) Two of the products of this factory can be used to make hydrochloric acid. Which are they?

(g) Brinetech have to buy their supplies of brine and sodium carbonate. What else do the company need to pay for that will be a major part of the cost of operating this factory?

(h) Brinetech sells most of its products to these buyers:

Sodium hydroxide — to a soap manufacturer
Chlorine   — to a PVC manufacturer
Hydrogen   — to a margarine manufacturer

(i) How does the soap manufacturer use the sodium hydroxide in making soap?
(ii) If the PVC manufacturer goes out of business, Brinetech must find a new market for its chlorine **at once**. Why must Brinetech do so?
Suggest another use for which Brinetech might sell its chlorine.

MEG

7 Four samples of tap water, **A, B, C,** and **D**, were collected from different parts of Britain. 20 cm^3 of sample **A** was placed in a test tube, soap solution was added 1 cm^3 at a time and the mixture was shaken. The addition of soap solution, with shaking, was continued until a permanent lather was obtained. The experiment was repeated using 20 cm^3 of a boiled sample of **A,** and also with 20 cm^3 of sample **A** to which sodium carbonate (washing soda) had been added

The experiments were repeated using samples **B, C** and **D**.
The results are shown in the table.

Water sample	Volume of soap solution needed to make a lather (cm^3)		
	before treatment	after boiling	after adding sodium carbonate
A	1	1	1
B	25	1	1
C	15	15	1
D	20	10	1

(a) Which of the tap water samples contains chemicals which cause
(i) temporary hardness only

(ii) permanent hardness only

(iii) both temporary and permanent hardness?

(b) Why are equal volumes of water used in these experiments?

(c) Calcium sulphate is a chemical which causes permanent hardness in water. How does it get into water by a naturally occurring process?

(d) State one advantage and one disadvantage of hard water.

(e) All four samples of tap water were found to contain small quantities of chloride ions.

(i) Describe and state the result of a test which might be used to show that chloride ions are present in the water.

(ii) Suggest the most likely reason for these ions being present in tap water.

NEA

8 (a) The first stage in the manufacture of sulphuric acid is the oxidation of sulphur dioxide to sulphur trioxide (SO_3) at 450°C with a vanadium(v) oxide catalyst.

(i) What do you understand by the term *catalyst*?

(ii) Of which group of elements in the Periodic Table is vanadium a member?

(iii) Construct an equation for the catalytic oxidation reaction described above.

(b) The exhaust systems of American cars are fitted with special catalysts to convert carbon monoxide and nitrogen monoxide into safer products.

$$2CO + 2NO \longrightarrow 2CO_2 + N_2$$

(i) Name an additional pollutant present in the exhaust fumes of British cars. How does it get there?

(ii) Explain briefly why carbon monoxide is dangerous.

(iii) Explain why carbon dioxide can also be considered an atmospheric pollutant.

(iv) How might nitrogen monoxide be formed in a car engine?

(v) From your knowledge of nitrogen monoxide, suggest one way in which its presence in the air might cause harmful effects.

(c) In hard water areas concentrated nitric acid is sometimes added to irrigation water in greenhouses to clear scale from the piping.

(i) What is the origin of the scale?

(ii) Write an equation to show how the scale is removed.

(iii) What is the other benefit of using nitric acid in the irrigation system?

SEG

9 (a) Sulphur dioxide is used for the manufacture of sulphuric acid.

(i) Give the chemical formula of sulphur dioxide.

(ii) Give **one** other important use of sulphur dioxide.

(b) A student warmed an excess of copper(II) oxide with dilute sulphuric acid and then filtered the mixture.

    (i) Name the salt formed by this reaction.
    (ii) Why was an excess of copper(II) oxide used?
    (iii) What was the reason for warming the mixture?
    (iv) Why was the mixture filtered?

(c) Why are some power stations a large source of sulphur dioxide pollution?

(d) What effect would sulphur dioxide pollution in the atmosphere have on
    (i) the pH of rain water;
    (ii) buildings or statues made of limestone?

(e) A student decided to test whether sulphur dioxide in air caused greater corrosion of iron by setting up the following apparatus.

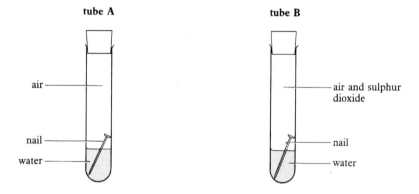

    (i) In **tube A** the nail forms rust (hydrated iron(III) oxide). Draw diagrams of **two** tubes the student could set up to show that neither water alone nor air alone causes the iron nail to rust.
    (ii) How could the student show that the nail in **tube B** corrodes faster than the nail in **tube A?**

SEG

8 *Metals and their reactions*

8.1 *Metals*

The pound in your pocket

At one time, coins were made out of silver and gold so they were actually worth the value they represented. However, the value of the metal of some coins soon became far more than the face value of the coin, and anyway, silver and gold were too soft and the coins wore away. Nowadays, coins are expected to last at least 40 years. Modern coins are made of alloys. One and two pence pieces are made of bronze (copper and tin) and silver coins are made of cupronickel (copper and nickel). At the Royal Mint, copper, nickel, manganese and zinc are melted together in an electric furnace at 1000°C to make pound coins. The alloy produced is much harder and more resistant to corrosion than any of the separate metals. To be sure that the alloy is of the correct composition, samples are examined by X-rays before the molten mixture is cast into long, thin slabs. These are pressed between a series of rollers at very high pressures and blanks (plain discs of metal) are stamped out with very hard tungsten carbide cutters.

Next, the blanks are heated to 700°C in a furnace and then cooled slowly. Any corrosion on the outside is cleaned off by soaking the blanks in sulphuric acid and then washing and drying them.

The design on the coins is stamped on. The coin blanks are pressed between the two dies at a pressure of 50 tonnes. This not only produces the design but raises a rim around the edge of the coin. The rim stops the faces of coins rubbing together so that they last even longer.

Metals are very useful materials. They can be mixed, squashed, heated up and softened and stamped and yet still remain strong.

- Why did the Bank of England replace the £1 note with a coin?
- Banks weigh bags of coins to see how much is in them. Why is it important that even old coins don't wear out too quickly?
- How much heavier is a 10p coin than a 5p coin, do you think? Check your guess on a balance.

Figure 1
How long do you think coins like this will last?

The position of metals in the periodic table

Figure 2 shows where metals are found in the periodic table. The properties of the metals change depending upon where they are in the table. Metals in Groups I and II are very reactive and not metal-like in their physical properties. Those in the Transition block are much heavier and harder. Metals in Groups III and IV often make compounds which have chemical properties rather like those of non-metals.

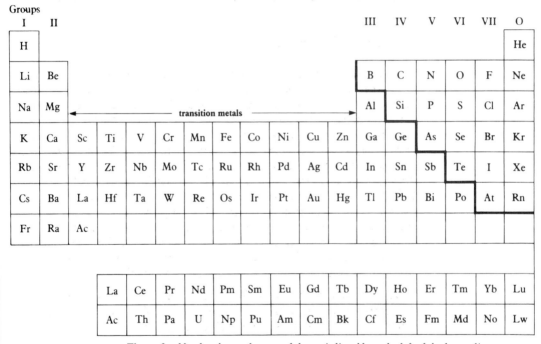

Figure 2 *Metals take up the part of the periodic table to the left of the heavy line.*

What makes metals special?

Most metals have a number of physical properties in common, but there are always exceptions.

1 Most metals are sonorous. They make a pleasant noise when hit, especially when they are made into a shape that vibrates like a bell or a gong.

2 Most metals are malleable and ductile. When they are hammered or

rolled, they can be made into thin sheets. Hot steel is rolled into steel plate for making car bodies and copper is rolled and stretched into very thin copper wire.

3 Metals are good conductors of heat and electricity. The thin lines on printed circuit boards are made of copper.

4 Most metals are shiny, but are easily corroded by air and water, although gold and silver are very unreactive and stay bright and uncorroded. The wires connecting a microchip to its base are made of gold because they are free from corrosion and conduct electricity very well.

5 Some metals are hard, for example iron, manganese and chromium. However, metals like sodium and potassium are very soft and can easily be cut with a knife.

6 Some metals are heavy, such as gold, lead and mercury. Others like magnesium and aluminium are light. Sodium and potassium are even lighter than water and float.

7 The 'odd man out' of the metals is mercury because it is a liquid. However, it has many of the other properties of metals. For example it is heavy, shiny and a good conductor of heat and electricity.

Figure 3
This gold beater is using a leather hammer to make very thin gold leaf. Why are the hammers made of leather?

A chemist's view of metals

When metals react to form ionic compounds, they form positive ions. Sodium, potassium and magnesium react readily with air, water and acids in each case forming positive ions.

sodium + water $\longrightarrow$ sodium hydroxide + hydrogen

$2Na(s) + 2H_2O(l) \longrightarrow 2NaOH(aq) + H_2(g)$

The oxides formed by most metals are called basic oxides, or bases. They neutralise acids forming salts.

magnesium oxide + sulphuric acid $\longrightarrow$ magnesium sulphate + water

$MgO(s) + H_2SO_4(aq) \longrightarrow MgSO_4(aq) + H_2O$

Internal structure

The atoms in metals are packed closely together in a regular arrangement. Figure 4 shows the three ways they can fit together. These structures give metals their properties.

Figure 4
Metal atoms are arranged in one of these three ways.

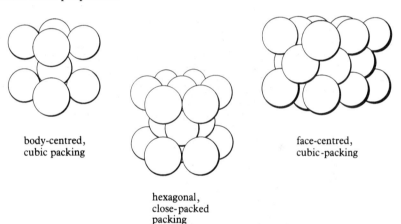

body-centred,
cubic packing

hexagonal,
close-packed
packing

face-centred,
cubic-packing

The electrons in the outside shells of the atoms can be delocalised into a 'sea' of electrons that are free to move through the metal. This enables metals to conduct heat and electricity well.

Layers of atoms can 'slip' over each other. This enables metals to be squashed and stretched. The hotter a metal is made, the easier this movement becomes. However, although the arrangement of atoms in a metal is regular, there are often small interruptions or 'dislocations' between some rows of atoms which stop the movement of atoms. This produces 'crystal grains' which can easily be seen in some metals. (Figure 5.) Blowing bubbles in soapy water illustrates how atoms pack together regularly and how breaks in the structure produce boundaries dividing the structure up into grains. The bubbles represent the atoms!

Steel can be made harder by heating it up and plunging it into water. This creates lots of tiny crystal grains within the metal, so stopping slip between the atoms and this makes the metal resistant to bending and stretching.

Figure 5
*The arrangements of
atoms in a metal can be
illustrated by using
bubbles. The breaks in
the structure are called
dislocations.*

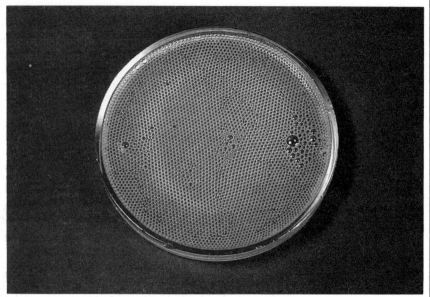

Alloys

Another way of making a metal stronger is by putting atoms of another metal into its structure. This stops the layers of atoms from moving so easily. Mixtures of metals like this are called **alloys**. Not only is the strength of the metal changed, but other properties such as electrical conductivity and resistance to corrosion are altered too.

Here are some examples.

1 Brass is a mixture of 60% copper and 40% zinc. Brass is harder than either copper or zinc. It doesn't corrode so easily and stays much shinier. Apart from being used for ornaments, brass is important for electrical connections and machine parts.

2 Bronze is made from copper and 2 or 3% of tin. It is used for 'copper' coins, metal for making guns, and parts of machines that take a heavy wear. It is much harder than either of its metals, and stronger than brass.

3 'Silver coins' such as 5p and 10p pieces are made from cupronickel, an alloy of 75% copper and 25% nickel. It is hard and resistant to corrosion, and much cheaper than real silver.

4 Solder is made from lead and tin. This alloy has a much lower melting point than either the lead or tin from which it is made and can be easily melted with a hot iron (Figure 6).

5 Duralumin is aluminium mixed with 4% of copper and 0.5% each of magnesium and silicon. It still retains the lightness of aluminium, but is much stronger. It is used for springs, non-magnetic tools and in places where sparks would be dangerous.

6 Carbon steel is a mixture of iron and carbon. The amount of carbon affects the properties of the alloy. Cast iron that comes straight from a

blast furnace always contains more than 5% of carbon. This makes it brittle. If the carbon content is reduced to about 1.5%, the alloy becomes very hard and strong. It is used for tools, girders, and items such as knives and scissors. Steel with less than 1.5% of carbon is called mild steel and is much softer. It can be rolled and pressed into shapes more easily. The steel for car bodies is made from mild steel.

7 Alloy steels contain other metals as well as carbon. Stainless steel has chromium and nickel added to it. Very hard steel for drills and screwdrivers contains tungsten. Steel for magnets has cobalt in it and titanium makes steel which can resist very high temperatures.

Figure 6
The solder has a low melting point and quickly becomes a liquid when heated.

Key points

- Metals are good conductors of electricity.
- Most metals form positive ions when they react and their oxides are generally basic.
- Atoms in metals are close packed. Small dislocations in their structure make them strong.
- Alloys are mixtures of metals which often have more useful properties than pure metals.

Quick questions

1 Give two physical properties and two chemical properties of sulphur that show it is not a metal.

2 Suggest a reason why electrical connections in undersea telephone cables are made with gold and not copper.

3 Gold is often made into an alloy with copper. How do you think this might affect the properties and cost of the gold?

8.2 *Extracting metals*

Metals from under the sea

The sea contains many million tonnes of metal compounds, but their solutions are so dilute that the cost and difficulty of extracting them far exceeds the value of the metals. One metal, magnesium, can be extracted from the sea economically, and hundreds of tonnes of magnesium are made each year by the electrolysis of sea water.

The other source of metals is rocks which are mined for their metal ores. However, a great deal of the surface of the Earth is under water, and so many sources of minerals are impossible to mine because they are under thousands of metres of water in the middle of the oceans. Deep sea research, using remote control pressurised vessels with television cameras, has shown that there are '**nodules**' — flat, round lumps weighing about a kilogram — containing manganese, nickel, cobalt and copper and other metals, on the sea bed in certain parts of the world. Collecting these nodules is expensive, but scientists think that remote control scoops connected to long suction pipes could be used to bring these metals to the surface economically. It would certainly cost less then mining in the middle of the oceans!

Ores and mining

Metals are generally quite reactive substances, so most of them are found in the ground as compounds called **ores**. Ores are oxides, sulphides, sulphates, carbonates and other compounds of metals.

Some metals, like gold and platinum, are very unreactive and are found **native**, as the pure element, but mixed up with rock. Many metals are not only expensive because they are rare, like gold and platinum, but because the mining and extracting processes add to costs as well. Ores that are found near the surface are obtained by **open cast mining**. The surface layer of soil and rock (called the **overburden**) is stripped away to expose the ore. This is then removed with big mechanical excavators. Afterwards, the overburden is replaced and the mine is landscaped to make it look as attractive as possible. None-the-less, open cast mining looks dreadful whilst it is in progress.

Ores further down in the ground are mined by **open pit mining**. The people and equipment in Figure 2 give some idea of the scale of things. Ores even deeper down in the ground are mined like coal.

Extracting metals from their ores

The way in which a metal is extracted from its ore depends upon how reactive it is. Very unreactive metals like gold are found native. Very reactive metals like potassium, are very strongly combined with the other

Figure 1
The crystals in this rock are gold. It takes 2/3 of a million tonnes of gold-bearing rock to produce 10 tonnes of gold.

Figure 2
160 tonnes of copper ore must be mined to extract one tonne of copper. What quantity of copper do you think has been extracted from this hole?

elements in their ores. Other metals like lead and copper are less strongly combined.

When a metal is removed from its ore, this is done by the process of **reduction**.

Oxide ores are heated in a blast furnace with coke. For example:

iron ore + carbon monoxide $\longrightarrow$ iron + carbon dioxide

$$Fe_2O_3(s) + 3CO(g) \longrightarrow 2Fe(l) + 3CO_2(g)$$

from the coke

Sulphide ores are *roasted* in retorts. For example:

lead sulphide + oxygen $\longrightarrow$ lead + sulphur dioxide

$$PbS(s) \quad + \quad O_2 \quad \longrightarrow \quad Pb(l) + \quad SO_2(g)$$
(air)

For very reactive metals like sodium, calcium, magnesium and aluminium, the reduction has to be carried out by **electrolysis** of the ore instead.

The production of aluminium

Aluminium is made by the electrolysis of the ore, **bauxite** ($Al_2O_3.3H_2O$). First of all, the ore is crushed and added to hot sodium hydroxide solution. The bauxite dissolves, but impurities of sand, iron oxide and titanium oxide do not. The mixture is filtered and some pure aluminium oxide is added to make the purified bauxite precipitate from the solution. The precipitate is dried.

Aluminium oxide has a melting point of 2040°C, so it is dissolved in molten **cryolite** (Na_3AlF_6) and a solution is formed at the much lower temperature of 1000°C. This is the electrolyte for the production of aluminium. Figure 3 shows the cell which is used.

Figure 3
Molten aluminium collects at the bottom of this electrolysis cell and is sucked out mechanically. The carbon anodes burn away in the hot oxygen and have to be replaced.

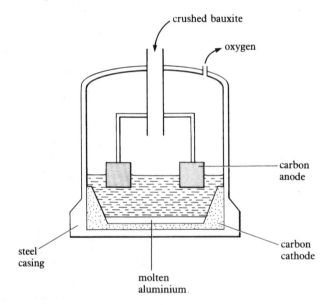

At the cathode, aluminium ions change into molten aluminium.

$$Al^{3+} + 3e \longrightarrow Al(l)$$

The aluminium sinks to the bottom of the cell and is siphoned off to be cast into small ingots.

At the anode, oxygen is produced.

$$2O^{2-} \longrightarrow O_2(g) + 4e^-$$

As the hot oxygen escapes, it burns away the graphite anode. The anode has to be replaced from time to time.

Making aluminium is an expensive process because 15 000 kWh of electricity are needed to make one tonne. Aluminium works are found where there is cheap hydroelectric electricity such as Scotland, Norway and Canada, or in countries where coal (for electricity) is cheap, such as Germany and Australia.

Iron and steel

Iron, and its alloy, steel, are the most-used metals in the World. 60 million tonnes are produced each year, of which 10 million tonnes are made in Great Britain.

Iron ore, iron(III) oxide, mined in Brazil, Canada, Sweden and UK, is mixed with **limestone** (calcium carbonate) and **coke** (made from coal). These starting materials are crushed and fed into the top of a **blast furnace**. (Figure 4.)

Figure 4
This blast furnace works non-stop. Ore, coke and limestone are fed in at the top and cast iron and slag are tapped off at the bottom. Why do you think it is never allowed to go out?

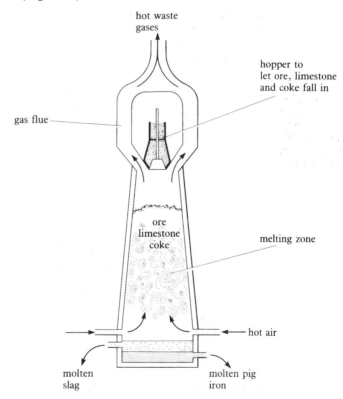

At the foot of the blast furnace, very **hot air** is blasted in, making the coke burn exothermically.

$$C(s) + O_2(g) \longrightarrow CO_2(g)$$
coke   carbon dioxide

The carbon dioxide is reduced by more coke to form **carbon monoxide**.

$$C(s) + CO_2(g) \longrightarrow 2CO_2(g)$$
carbon monoxide

Carbon monoxide is a **reducing agent** and it reduces the iron(III) oxide to molten iron. The molten iron runs down to the bottom of the furnace.

$$3CO(g) + Fe_2O_3(s) \longrightarrow 2Fe(l) + 3CO_2(g)$$
iron(III) oxide

At the same time, the limestone decomposes.

$$CaCO_3(s) \longrightarrow CaO(s) + CO_2(g)$$
limestone   calcium   carbon dioxide
oxide

Calcium oxide is a basic oxide and reacts with any sand in the ore to form **slag**.

$$CaO(s) + SiO_2 \longrightarrow CaSiO_3(l)$$
calcium   sand   slag
oxide

Molten slag runs to the bottom of the furnace and floats on the molten iron.

As the extraction goes on, more ore, limestone and coke are put into the top of the furnace and slag and iron are tapped off at the bottom.

The gases that come out of the furnace — nitrogen, carbon dioxide, carbon monoxide, steam, sulphur dioxide and oxygen — are hot. Rather than just waste the heat the gases are used to heat the air which is blown in at the base of the furnace to heat the reaction.

Slag is of little value and is used for such things as road making. The iron formed, called **cast iron** or pig iron, is brittle because of the large amount of carbon and other impurities dissolved in it. Some of it is used to make things like manhole covers and car engine blocks, but most of it is used in **steel** production.

To make steel, the iron must first of all be purified and then small, exact amounts of carbon and metals are added. In the **basic oxygen** process, molten iron from the blast furnace is mixed with 30% scrap iron in a **steel converter**. Oxygen is blown into the molten mixture and impurities of carbon and phosphorus are blown out as gases. Limestone is added and slag is formed, with the silicon dioxide still in the iron. The slag is scraped off the surface of the molten metal. Before the liquid steel is poured out, it is analysed for carbon content and metals are added in the right amount if **alloy steel** is to be made. Finally, the liquid steel is poured into ingots ready to be forged with giant hammers or continuously cast into long girders. It may, instead, be rolled into long, flat strips for such things as car bodies and baked bean cans.

Copper

Copper sulphide ore comes from Canada, Chile, Zambia and the USA. The ore is first of all crushed and separated from unwanted rock by **froth flotation**. Ore and water are stirred vigorously with chemicals which make a froth. The particles of ore cling to the froth bubbles while the rock sinks.

After drying, the ore is **smelted** by roasting it with sand and air in a furnace. Slag is formed, sulphur dioxide is driven off and a purified form of copper sulphide called **matte** is left.

The matte is again smelted and impure copper called **blister copper** is left. This is refined by heating it in a stream of air before it is **electro-refined**. The impure copper is made the anode in an electrolysis tank containing copper sulphate solution. A thin, very pure copper cathode is used which gets fatter as copper is plated onto it during the electrolysis. (Figure 5.) Copper obtained in this way is 99.99% pure.

Figure 5
Copper for electric wires needs to be very pure. Pure copper is made by the process of electro-refining. Compare it with the electroplating process in section 5.5.

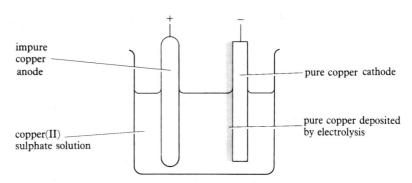

Extract your own metals

Small amounts of copper and lead can be made by heating their ores with carbon.

1 Use copper(II) oxide as copper ore. Mix small amounts of black, copper(II) oxide and charcoal powder. Put the mixture into a hard glass test tube.

2 Heat the mixture with a medium flame and as soon as the mixture starts to glow, take the test tube out of the flame and see if the glow continues on its own. Is the reaction exothermic?

3 As soon as the glow fades, allow the tube to cool a little and then look at the products through the wall of the test tube. If you see an orange/brown colour, that's the copper.

$$2CuO(s) + C(s) \longrightarrow 2Cu(s) + CO_2(g)$$

Why is it important not to tip the hot copper out into the air? What would happen to it?

Now use yellow lead(II) oxide as lead ore.

1 Mix together lead oxide and charcoal powder on a tin lid or a piece of heat-proof paper.

2 Heat the mixture as strongly as possible for several minutes. Any lead will be seen as tiny molten balls.

$$2PbO(s) + C(s) \longrightarrow 2Pb(l) + CO_2(g)$$

The importance of metals

Here is a list of some of the more important metals with some of the uses of the metals and their compounds.

Metal	Use
aluminium	kitchen foil, milk bottle-tops, window frames, alloys, aircraft, buses, pots and pans.
copper	electric wiring, water pipes, brass, coins.
chromium	stainless steel, plating, compounds for dyeing.
gold	jewelry, microchip wiring.
iron	steel of all sorts, ships, cars, girders for buildings, food-cans.
lead	batteries, petrol additives, solder.
magnesium	alloys, fireworks, flares.
molybdenum	steel, lubricants.
nickel	stainless steel, plating.
platinum	catalysts, thermocouples.
tin	food cans, solder, bronze, pewter.
silver	jewelry, knives and forks.
tungsten	aerospace alloys, filaments for bulbs, drill tips, tools, white paint.
zinc	dry-cell batteries, galvanising, brass, ointments.

- What property of gold makes it so valuable for jewelry?
- What uses of tungsten tell you that the metal has a high melting point?
- Find the two metals that are used with iron to make stainless steel.
- Find out (look on a tin lid or tube) what compound of zinc is used in ointments.

Finite resources

Metals are **finite resources**. Once we have mined and processed all the metal ores in the ground, there will be no more left. New ores are not being formed. The rate at which we are using up our resources is alarming. Figure 6 shows an estimate of how long our resources of metals (with coal, oil and gas for comparison) will last if we continue to use them at our present rate. It will be increasingly important to conserve our metals. Many metals can be **recycled**, but often the cost of recycling is greater than the cost of processing new ore — at the moment! However, scrap iron and steel is already added to new iron from blast furnaces to make steel. Lead is recovered from old car batteries and copper is reclaimed from electric wiring. A lot of aluminium is recovered from drink cans, kitchen foil and milk bottle-tops. Unlike other metals, it is cheaper to reprocess aluminium than it is to extract it from its ore.

Figure 6
This is a rough estimate of how long some of our resources will last if we continue to use them at the present rate. How old will you be when there is no copper left?

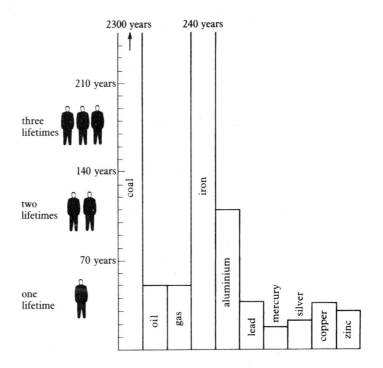

Key points

- Most metals are found in the ground as ores and have to be mined.
- Apart from native metals, metals have to be extracted from their ores by reduction.
- Aluminium is extracted from bauxite by electrolysis.
- Iron is extracted from its ore by reduction with coke in a blast furnace, and is then made into steel.
- Copper is obtained from its ore by roasting it, and it is then purified by electrolysis.
- Metals are finite resources. They cannot be replaced when the Earth's supply has been used up.
- Some metals can be economically recycled.

Quick questions

1 Which methods would you use to extract (a) potassium (b) lead from their ores? Explain your choices.

2 Sketch the apparatus you might try to use in the laboratory to extract sodium from sodium chloride by electrolysis. Explain what would go on at each electrode.

3 Why is the smelting of copper likely to be an environmental hazard?

4 Explain why limestone is an important ingredient in the manufacture of iron and steel.

8.3 *Reacting metals*

The strange behaviour of aluminium

Aluminium is a useful metal because it is strong and light, and hardly corrodes at all. Window frames and greenhouses made of aluminium last without rusting away. They don't even need to be painted. When aluminium is exposed to air, a very thin coating of aluminium oxide quickly forms on its surface and protects the metal underneath. Air, water and even acid cannot get through the oxide layer.

Aluminium pieces for window frames are made in long strips, by a process called **extrusion**. Extrusion is like squeezing toothpaste out of a tube. An ingot of aluminium is heated to 500°C and is forced through a small piece of steel with the correct shape cut in it. The long strip of aluminium comes out of the 'nozzle' and can be cut off in convenient lengths.

In a similar way, hot aluminium can be rolled between heated rollers to make very thin sheets of aluminium foil. (See Figure 1.)

The Activity Series

Metals are put into a table, called the **Activity Series**. The most reactive metal is at the top of the series. The least reactive metal goes at the bottom. The activity table is useful as a reminder about a metal's reactions and for predicting unknown reactions.

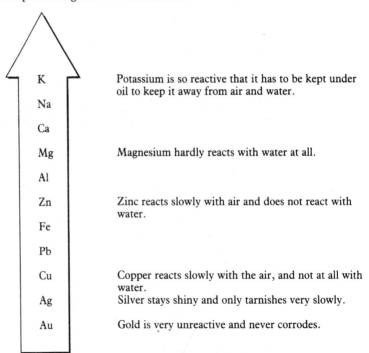

K   Potassium is so reactive that it has to be kept under oil to keep it away from air and water.

Na

Ca

Mg   Magnesium hardly reacts with water at all.

Al

Zn   Zinc reacts slowly with air and does not react with water.

Fe

Pb

Cu   Copper reacts slowly with the air, and not at all with water.

Ag   Silver stays shiny and only tarnishes very slowly.

Au   Gold is very unreactive and never corrodes.

Figure 1
Hot aluminium can be rolled between heated rollers to make very thin foil. A slab of aluminium weighing 10 tonnes can end up as thousands of metres of foil just 0.015 mm thick.

Reaction with oxygen

One way of putting metals into their correct places in the Activity Series is by studying the way they react with oxygen. Some metals burn in air — and more brightly in oxygen. Magnesium burns brightly in air and is used for flares and fireworks and it is small particles of iron that make sparklers sparkle.

Some metals do not burn in air but form oxides on their surface. Zinc and lead are used on roofs because once a surface coating of oxide is formed, the metal underneath is protected. Copper-containing coins slowly go brown and finally black as copper oxide is formed. Gold and silver are used as jewelry metals because they do not form oxides in air or oxygen.

Here is a summary of the way metals react with the oxygen in the air.

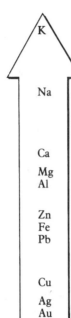

These metals burn brightly in air to produce oxides. Their oxides dissolve in water to form alkalis. Calcium is a good example:

$$2Ca(s) + O_2(g) \longrightarrow 2CaO(s)$$

Calcium burns to form calcium oxide. Calcium oxide dissolves exothermically in water to form calcium hydroxide solution.

$$CaO(s) + H_2O(1) \longrightarrow Ca(OH)_2(aq)$$

Sodium reacts like this, but more vigorously.

These metals all burn in air (magnesium brightly, aluminium less so, zinc slowly) to form oxides that do not dissolve in water. For example:

$$2Zn(s) + O_2(g) \longrightarrow 2ZnO(s)$$

These metals will not burn in air. Instead they oxidise and become coated with a layer of oxide, which does not dissolve in water. For example:

$$2Cu(s) + O_2(g) \longrightarrow 2CuO(s)$$

These metals do not react with oxygen at all.

Reactions of metals with water

Metals react in different ways with water or steam according to where they are in the activity series.

These metals react very vigorously with water, releasing hydrogen and leaving behind an alkaline solution. For example

$$2Na(s) + 2H_2O(1) \longrightarrow 2NaOH(aq) + H_2(g)$$
sodium   sodium hydroxide

Magnesium reacts very slowly water. The reaction very soon slows up however, because a coating of magnesium hydroxide is formed which is not very soluble.

Magnesium reacts very vigorously with steam. (Figure 2).

$$Mg(s) + H_2O(g) \longrightarrow MgO(s) + H_2(g)$$
magnesium   magnesium oxide

These metals only react with steam. They become less and less reactive lower down the series.

These metals have no reaction at all with water or steam.

Figure 2
Why couldn't firemen use water to put out a fire at a magnesium factory? What might happen if they tried?

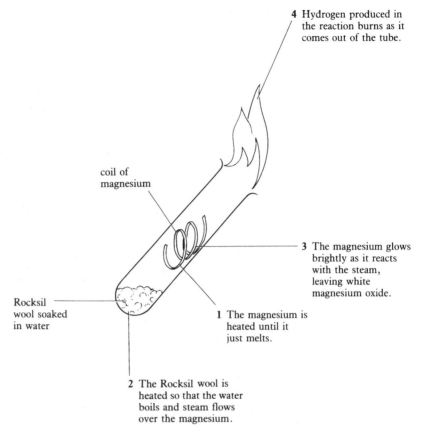

4 Hydrogen produced in the reaction burns as it comes out of the tube.

coil of magnesium

3 The magnesium glows brightly as it reacts with the steam, leaving white magnesium oxide.

Rocksil wool soaked in water

1 The magnesium is heated until it just melts.

2 The Rocksil wool is heated so that the water boils and steam flows over the magnesium.

Reaction of metals with dilute acids

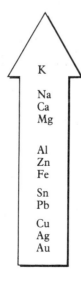

K

Na
Ca
Mg

Al
Zn
Fe

Sn
Pb

Cu
Ag
Au

These metals react with dilute sulphuric acid and dilute hydrochloric acid to form salts and hydrogen. The reactions of sodium and potassium are dangerous because so much heat energy is evolved. Zinc is, therefore, generally used in the laboratory to make hydrogen by reacting it with acid.

$$Zn(s) + H_2SO_4(aq) \longrightarrow H_2(g) + ZnSO_4(aq)$$
zinc   zinc
                                              sulphate

Tin only reacts very slowly and the reaction of lead soon stops because insoluble salts are formed.

These metals do not react at all with dilute sulphuric or hydrochloric acid.

Rusting

Some forms of **corrosion** are very useful. The coating of oxide on the surface of aluminium protects the metal below. Metals like lead and zinc are used for weather-proofing, because once oxidised by air and water, they are protected from further corrosion.

But the corrosion of iron, **rust**, is different. Rust is iron(III) oxide and is caused by the reaction of iron with water and air. Both air and water are needed. Once rusting has started, it is very difficult to stop, and the corrosion can weaken or even completely eat away the metal. To prevent rusting would seem a simple matter. Either remove the water or the air causing the rusting!

Methods of preventing rusting

1 keeping the iron dry with **silica gel** or **calcium chloride**.
2 keeping the iron air free in a **vacuum**.
   (1 and 2 are not always practical.)
3 **painting** the iron.
4 covering the iron with **oil** or **grease**. (This is fine for machines.)
5 **electroplating** the iron with chromium or silver. This is **expensive**.
6 coating the iron with molten **plastic**.
7 **galvanising** the iron.
8 **tin plating** the iron.
9 using **sacrificial corrosion**.

Figure 3
This bridge will rust easily if it is not protected. The steel is coated with zinc and then paint to keep air and water out. Why is it coated with zinc first?

Galvanising

Objects like buckets, nails for outside use, corrugated iron and water tanks are dipped into molten zinc as soon as they are made. The coating of zinc protects the iron from air and water. When the zinc corrodes, it forms a thin layer of white zinc oxide and then corrosion stops.

But, buckets get dented and chipped and the iron underneath becomes exposed. Even then, the zinc continues to protect the iron. Because zinc is higher in the activity series than iron (it is more reactive), the zinc continues to corrode instead of the iron, even though the iron is exposed.

Tinning

Food cans are made of steel, which would go rusty when foods containing water are put in. The inside of the can cannot be galvanised, because the zinc would react with the food contents too, especially if it were acidic, like fruit. Instead, the can is plated with a thin layer of tin, which is very unreactive. This layer of tin is in turn covered with a coating of lacquer or plastic to complete the protection. This works well unless the tin gets dented. If the lacquer and tin layers are broken, the iron becomes exposed. The tin does not protect the iron because the iron is the more reactive metal.

Figure 4 *Even if the zinc coating on this bucked is damaged and air and water get to the iron, it is still protected because the zinc corrodes instead of the iron.*

Figure 5 *This could be dangerous. What might have happened if the coating inside this tin is damaged?*

Sacrificial corrosion

Ships are made of steel, so blocks of zinc are bolted to the underside of ships so that they corrode and dissolve instead of the steel.

Figure 6
Bars of zinc attached to a ship's hull to protect it from corrosion.

The best method of rust protection

In this experiment, iron nails are put into test tubes under different conditions. (Figure 7.)

1 The first nail is put into a test tube of tap water. This one should rust. It is called the **control**.
2 The second nail is put into a test tube containing dry calcium chloride and stoppered tightly so no damp air can get in.
3 Nail number three goes into some water that has been boiled to drive

out all dissolved air. The top of the test tube is sealed with molten candle wax to keep air out.

4 The fourth nail is painted and left in tap water.

5 The next nail is coated in vaseline and left in tap water.

6 A piece of magnesium ribbon is wound around the nail and it is left in tap water. Magnesium is more reactive than iron and should have the same effect as galvanising or sacrificial corrosion.

7 A thin strip of copper is wound around the last nail. Copper is less reactive than iron.

Figure 7
The rust test.

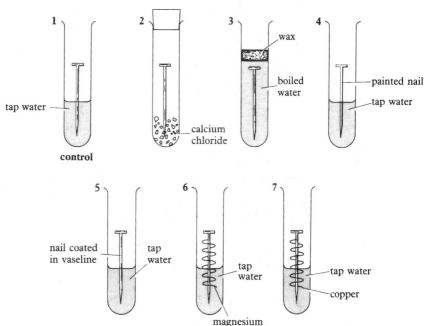

All the nails are left for a day or two, and the results studied.

- Which nails will not rust?
- Why is it important to have a control tube?
- Does tube 2 contain any water? Does tube 3 contain any air? Explain your answers.
- What will happen in tube 6?
- Why will the nail in tube 7 still rust?

Key points

- Aluminium does not corrode because of its protective coating of oxide.
- Metals can be put in an Activity Series in the order of their reactivity with air, water and dilute acids.
- Rusting is caused by air and water together.
- Rusting can be prevented if either air or water is excluded.
- Galvanised iron is coated with zinc, which corrodes instead of the iron.
- Food cans are tin plated to prevent the steel rusting.

Quick questions

1 Look around your home and try to see how many things you can find that could be made by the process of extrusion. Start by looking where the walls meet the floor, and at the tops of the curtains.

2 An unknown metal isn't kept under oil, but it does form a thin layer of oxide. When put into water, it fizzes very slowly, and reacts quite vigorously with dilute acids. It protects iron from rusting by sacrificial corrosion. What is the metal and why wouldn't it be sensible to make pipes for a steam boiler from this metal?

3 A metal burns brightly in air and fizzes in water.
   (a) What effect would the solution formed have on litmus paper? Explain your answer.
   (b) This metal dissolves very exothermically and rapidly in dilute hydrochloric acid, but very slowly in dilute sulphuric acid. Why?

8.4 *Displacement and redox reactions*

New clothes for the Statue

Just over a hundred years ago, the people of France presented the USA with the Statue of Liberty. When he designed it, Alexandre Eiffel (of Eiffel Tower fame), had to know his activity series. The outer shape of the statue was made of 2.5 mm thick sheets of copper, hammered to the correct shape. To give it strength, the inside of the statue consisted of a wrought iron skeleton framework. But, here is where the chemistry started.

Iron is higher than copper in the activity series, so when the metals became wet with salty water during stormy weather, a chemical reaction would take place. The copper, iron and salty water would form an electrical cell and the iron would corrode rapidly. (More about this in the chapter that follows.) Also, in hot weather, the copper and iron would expand by different amounts, so they had to be able to move independently of each other. The copper couldn't be fixed rigidly to the iron skeleton. To overcome this problem, the copper outer casing was connected to the iron by strips of copper covered in asbestos and shellac. These strips allowed the copper and iron to move freely, and the asbestos and shellac made an insulating layer between them. Unfortunately, this insulating layer gradually became worn away with the movement, and the iron corroded so much, that it was necessary to do extensive repairs.

When the statue was reopened in 1986, much of the original iron had been replaced by stainless steel coated with a special anti-rust composition developed at the Kennedy Space Centre. The asbestos and shellac had been replaced by **Teflon** (the material used to coat non-stick saucepans) which is a tough plastic with a very low friction. It allows the copper to move freely on the steel skeleton.

One form of corrosion had to be retained however. Copper reacts with carbon dioxide and water in the atmosphere to form an attractive green coating. Where new rivets had been put in during the rebuilding, the new shiny metal had to be reacted with chemicals to produce this green colour artificially.

Figure 1
This well-known statue is made of copper and iron and stands in a very corrosive environment.

Displacement reactions

Silver whiskers

1 Wind a coil of copper wire around a pencil.
2 Half fill a test tube with silver nitrate solution and carefully lower the copper coil into it.
3 Be patient. Leave the test tube for an hour without disturbing it. The grey whiskers that grow on the copper are crystals of silver.
4 Repeat the experiment with a thin strip of clean zinc in lead nitrate solution. This time, the sparkling crystals are lead. (Figure 2.)

Figure 2
Silver and lead crystals can be made by displacement.

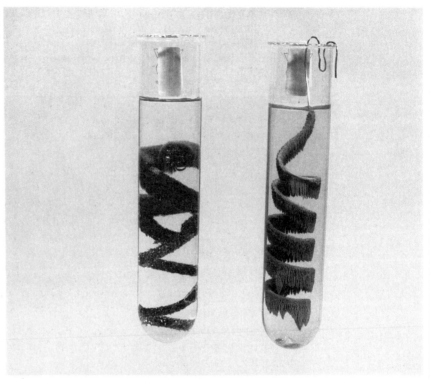

● Where is copper in relation to silver in the activity series?
● Is zinc higher or lower than lead in the activity series?

What is happening in displacement reactions

The silver and lead crystals in the experiment above have been made by **displacement reactions**.

$$Cu(s) + 2Ag(NO_3)_2(aq) \longrightarrow 2Ag(s) + Cu(NO_3)_2(aq)$$
    silver nitrate   copper(II) nitrate

Copper is higher in the activity series than silver. It reacts (or forms ions) more easily than silver. The copper atoms turn into copper ions and make the silver ions change into silver atoms.

The same thing happens in the second reaction in the experiment because zinc is higher in the activity series than lead.

$$Zn(s) + Pb(NO_3)_2(aq) \longrightarrow Pb(s) + Zn(NO_3)_2(aq)$$
    lead nitrate   zinc(II) nitrate

If a metal is put into a solution of the salt of another metal, which is lower in the activity series, then the lower metal will be displaced by the higher metal. Reactions of this type are called displacement reactions.

Examples of displacement reactions

Here are some more pairs of metals and solutions where displacement will take place.

$$Mg(s) + \quad CuSO_4(aq) \quad \longrightarrow Cu(s) + \quad MgSO_4(aq)$$
$$\text{copper(II) sulphate} \qquad\qquad \text{magnesium sulphate}$$

As the reaction takes place, a brown coating of copper appears on the surface of the magnesium and the blue colour of the copper(II) sulphate fades as colourless magnesium sulphate is formed. The magnesium has displaced the copper.

$$Fe(s) + \quad Pb(NO_3)_2(aq) \longrightarrow Pb(s) + \quad Fe(NO_3)_2(aq)$$
$$\text{lead(II) nitrate} \qquad\qquad \text{iron(II) sulphate}$$

Silver-grey lead is formed on the surface of the iron and the solution turns pale green as iron(II) sulphate is formed. The iron has displaced the lead.

Redox with metals

When a compound has oxygen taken away from it by another compound or element, a **reduction** reaction has taken place. Reducing agents which have been mentioned so far are hydrogen and carbon monoxide.

If copper(II) oxide is heated in a stream of hydrogen, the hydrogen reduces the copper(II) oxide by removing its oxygen. The hydrogen changes into steam and copper is left.

$$CuO(s) \quad + \quad H_2(g) \quad \longrightarrow Cu(s) + H_2O(g)$$
$$\text{oxidising agent} \quad \text{reducing agent}$$
$$\text{(this is reduced)} \quad \text{(this is oxidised)}$$

Hydrogen is the reducing agent because it takes oxygen away from the copper(II) oxide. The copper(II) oxide is reduced. At the same time, copper(II) oxide is the oxidising agent because it gives oxygen to the hydrogen, and the hydrogen is oxidised.

Reduction and oxidation are both taking place at the same time. The reaction is called a **redox reaction**.

Here is a redox reaction that takes place inside the blast furnace.

$$3CO(g) \quad + \quad Fe_2O_3(s) \quad \longrightarrow 3CO_2(g) + 2Fe(l)$$
$$\text{reducing agent} \quad \text{oxidising agent}$$
$$\text{(oxidised)} \qquad \text{(reduced)}$$

> In a redox reaction, the reducing agent takes in oxygen and is itself oxidised. The oxidising agent gives out oxygen and is itself reduced.

Like hydrogen and carbon monoxide, metals can be reducing agents too, when they react with oxides of other metals lower in the activity series. For example, chromium is manufactured by heating a mixture of aluminium powder with chromium(III) oxide.

$$2Al(s) + \quad Cr_2O_3(s) \quad \longrightarrow 2Cr(s) + \quad Al_2O_3(s)$$
$$\text{chromium(III) oxide} \qquad\qquad \text{aluminium oxide}$$

Aluminium is higher in the activity series than chromium, so aluminium reduces chromium(III) oxide and changes into aluminium oxide. Molten chromium is formed.

A similar reaction is used for making small amounts of molten iron for welding railway lines together. (Figure 3.) A mixture of iron(III) oxide and aluminium is ignited. The mixture burns very exothermically and the more reactive aluminium reduces the iron(III) oxide to molten iron. This is called the **thermit reaction**.

$$2Al(s) + \quad Fe_2O_3(s) \quad \longrightarrow \quad 2Fe(l) + \quad Al_2O_3(s)$$
iron(III) oxide   aluminium oxide

Figure 3
This redox reaction is very exothermic. Aluminium is reducing iron(III) oxide to make molten iron which welds together railway lines.

Try redox for your self

1 Make a mixture of one spatula-full of zinc powder and one spatula-full of copper(III) oxide powder.
2 Make the mixture into a small pile on a tin lid or a piece of heat proof paper on a tripod.
3 Heat the mixture from above with the hottest part of a Bunsen flame.
4 As soon as a reaction starts, remove the flame and see if the reaction carries on on its own. Is the reaction exothermic?
5 When the reaction has finished, tip the products into some dilute sulphuric acid.
6 Warm and stir the acid. Any particles of red/brown copper will sink to the bottom.

Redox without oxygen

So far, reduction and oxidation reactions have involved oxygen. Other reactions, however, can be redox reactions too.

The displacement reaction between magnesium and copper(II) sulphate solution can be split into two halves:

$$Mg(s) + CuSO_4(aq) \longrightarrow Cu(s) + MgSO_4(aq)$$

can be written as

$$Mg(s) \longrightarrow 2e^- + Mg^{2+}(aq)$$

and

$$2e^- + Cu^{2+}(aq) \longrightarrow Cu(s)$$

or, when added together

$$Mg(s) + Cu^{2+}(aq) \longrightarrow Cu(s) + Mg^{2+}(aq)$$

The sulphate ions have to be there, but they are not part of the main reaction. They are called **spectator ions**, and can be left out.

The magnesium atoms have been oxidised because they have changed from atoms into ions and have lost electrons.

$$Mg(s) \longrightarrow Mg^{2+}(aq) + 2e^-$$

Oxidation Is Loss of electrons.

The copper ions have been reduced because they have changed from ions to atoms and have gained electrons.

$$Cu^{2+}(aq) + 2e^- \longrightarrow Cu(s)$$

Reduction Is Gain of electrons.

Figure 4 shows you a way of remembering the definitions of oxidation and reduction.

Figure 4
An aid to the memory.

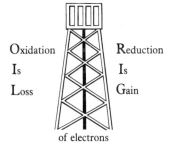

Oxidation   Reduction
Is   Is
Loss   Gain
of electrons

Here is another redox displacement reaction. Copper displaces silver metal from a solution of silver nitrate.

$$Cu(s) + 2AgNO_3(aq) \longrightarrow 2Ag(s) + Cu(NO_3)_2(aq)$$
silver nitrate   copper(II) nitrate

Or

$$Cu(s) + 2Ag^+(aq) \longrightarrow 2Ag(s) + Cu^{2+}(aq)$$

The copper atoms have been oxidised because they have lost electrons.

$$Cu(s) \longrightarrow Cu^{2+}(aq) + 2e^- \qquad OIL$$

The silver ions have been reduced because they have gained electrons.

$$2Ag^+(aq) + 2e^- \longrightarrow 2Ag(s) \qquad RIG$$

Key points

- If a metal is put into the solution of a compound of a less reactive metal, then the more reactive metal will displace the less reactive metal.
- Reduction is loss of oxygen.
- Reactions in which reduction and oxidation take place at the same time are called redox reactions.
- Reduction and oxidation can also be explained in terms of electrons. Oxidation is loss of electrons: reduction is gain of electrons.

Quick questions

1 For each of the following, complete the word equation, **only if** a reaction takes place:
   (a) iron + copper oxide $\longrightarrow$
   (b) calcium + zinc oxide $\longrightarrow$
   (c) lead + aluminium sulphate solution $\longrightarrow$
   (d) magnesium + tin(II) chloride solution $\longrightarrow$
   (e) lead + iron(III) oxide $\longrightarrow$

⋆ 2 Name the spectator ion in question 1(c).

⋆ 3 Explain why reactions 1(a) and 1(d) are redox reactions.

⋆ 4 Explain why this reaction is a redox reaction:

$$Mg(s) + 2HCl(aq) \longrightarrow MgCl(aq) + H_2(g)$$

8.5 *Families of metals*

The dangers of salt

Crisps, peanuts, Chinese take-aways, packet soups, baked beans, . . . all have one thing in common. They have lots of salt in them. Even nonprocessed food like potatoes, rice, stewed steak, cabbage, . . . seem to taste better if salt — sodium chloride — is sprinkled on them. Salt seems to bring out the flavour of some foods. We need a certain amount of salt, but only 0.2 g a day. An awful lot of people eat more than that, for the average British intake is 20 g a day!

Before fridges and freezers were invented, many foods were preserved in salt, and even today, salt is still the essential ingredient in many preserved foods. We don't need to eat as much salt as we do. We just seem to like it!

Too much sodium chloride is not only unnecessary but dangerous. Too many sodium ions lead to high blood pressure and that in turn makes some people more likely to have heart attacks, strokes, and kidney damage. So, leave out the salt (but don't forget the fat, sugar, alcohol, nicotine . . . as well!)

Figure 1
Do we need this salt, or do the beans just seem to taste better with it?

Group I the alkali metals

The metals in Group I of the periodic table are all grouped together because of one thing — they all have one electron in their outside shell.

This one electron makes them all very similar in their properties and reactions. When they react, they all do so in the same way, losing the one electron in the outer shell and forming a positive ion.

lithium   Li $\longrightarrow$ Li$^+$ + e$^-$
sodium   Na $\longrightarrow$ Na$^+$ + e$^-$
potassium   K $\longrightarrow$ K$^+$ + e$^-$
rubidium   Rb $\longrightarrow$ Rb$^+$ + e$^-$
caesium   Cs $\longrightarrow$ Cs$^+$ + e$^-$

Here are some of the properties and reactions Group I metals have in common.

1 They are all shiny on the inside, but heavily corroded on the outside. When they are cut and exposed to the air, they corrode and go black in seconds. The alkali metals are all kept under oil to prevent them from corroding away.

2 They are all soft enough to be cut with a knife. Lithium is the hardest and, as you go down the group, the metals become softer. Potassium is as soft as cheese.

3 They all have low melting points.
lithium   186°C
sodium   98°C
potassium   63°C
rubidium   39°C
caesium   30°C

4 They are all lighter than water, and so float.

5 They all burn with coloured flames and form oxides which form alkaline solutions in water. Hence their name.

lithium   : crimson flame
sodium   : yellow
potassium: lilac
rubidium   : blue/red
caesium   : blue/red

6 They all react vigorously with water, releasing hydrogen and dissolving to form an alkali. The vigour of the reaction increases as you go down the group.

Lithium floats, fizzes, dissolves and leaves lithium hydroxide.

$$2Li(s) + 2H_2O(l) \longrightarrow 2LiOH(aq) + H_2(g)$$
lithium
hydroxide

Sodium floats, rushes around, melts, fizzes and leaves sodium hydroxide solution.

$$2Na(s) + 2H_2O(l) \longrightarrow 2NaOH(aq) + H_2(g)$$
sodium
hydroxide

Potassium floats, rushes around, fizzes, melts, catches fire, spits and dissolves.

$$2K(s) + 2H_2O(l) \longrightarrow 2KOH(aq) + H_2(g)$$
$$\text{potassium}$$
$$\text{hydroxide}$$

Rubidium and **caesium** release so much energy that their reactions tend to be explosive.

Group II the alkaline earths

The metals in Group II are all in the group because they have two electrons in their outer shells. When they react, they lose two electrons and form ions.

beryllium	Be	This element doesn't behave like the rest
magnesium	Mg $\longrightarrow$ Mg^{2+} + 2e$^-$	
calcium	Ca $\longrightarrow$ Ca^{2+} + 2e$^-$	
strontium	Sr $\longrightarrow$ Sr^{2+} + 2e$^-$	
barium	Ba $\longrightarrow$ Ba^{2+} + 2e$^-$	

Because they all lose 2 electrons when they react the metals in Group II have very similar properties and reactions.

1 The metals corrode, but much more slowly than the Group I metals. Only barium is generally kept under oil.

2 They are much harder than the Group I metals. Only barium is soft enough to be cut with a knife.

3 None of them float on water.

4 All except for magnesium burn with a coloured flame.

      magnesium : brilliant white flame
      calcium   : brick red
      strontium : crimson
      barium   : apple green

The oxides they form become more alkaline as you go down the group.

5 The Group II metals are much less reactive in water than those in Group I.

Magnesium only just reacts with cold water, producing one or two bubbles. In hot water it fizzes, but the reaction soon stops as an insoluble coating of magnesium oxide builds up on the outside.
Calcium reacts more quickly, but leaves behind a cloudy suspension because calcium hydroxide is not very soluble. It does dissolve enough, however, to make the solution alkaline.
Barium fizzes rather like lithium and reacts much more quickly than calcium. It leaves a clear, alkaline solution because barium hydroxide is quite soluble in water.

Here are the equations for the reactions.

$$Mg(s) + 2H_2O(l) \longrightarrow Mg(OH)_2 + H_2(g)$$
$$Ca(s) + 2H_2O(l) \longrightarrow Ca(OH)_2(aq) + H_2(g)$$
$$Ba(s) + 2H_2O(l) \longrightarrow Ba(OH)_2(aq) + H_2(g)$$

Investigating Group II sulphates

1 Put small amounts of magnesium chloride, calcium chloride and barium chloride solutions into separate test tubes.
2 To each solution in turn, add drops of dilute sulphuric acid, first of all slowly, and then to excess.
3 Describe carefully what you see and write equations for any ionic precipitation reactions that have taken place.

● Are all Group II sulphates soluble?
● How does their solubility change as you go down the Group?

The transition metals

The transition metals occupy the central block of the periodic table. They have many properties in common.

1 They are all heavy compared with the metals in Groups I, II and III.

Metal	Number of times heavier than water
chromium	7.2
iron	7.9
copper	8.9
silver	10.5
mercury	13.5
gold	19.3
platinum	21.5
osmium	22.6

2 Some transition metals are very hard. Iron, chromium, manganese and titanium are good examples.

3 Some transition metals have very high melting points.

titanium   1677°C
manganese 1244°C
iron   1539°C

4 Some transition metals are very resistant to corrosion and stay bright and shiny. Good examples are chromium, nickel, silver, gold and platinum.

5 Transition metals are good conductors of heat and electricity.

6 Several transition metals or their compounds make good catalysts. Examples are vanadium(V) oxide in the Contact Process for making sulphuric acid; iron in the Haber process for making ammonia; manganese(IV) oxide for the decomposition of hydrogen peroxide; nickel for turning natural gas into hydrogen and other chemicals.

7 Many transition metals can form more than one ion.

Iron can form Fe^{2+} and Fe^{3+}
Copper can form Cu^+ and Cu^{2+}

8 Many transition metal compounds are brightly coloured. Here are some of them.

potassium manganate(VII)	$KMnO_4$	purple
iron(II) sulphate	$FeSO_4$	pale green
iron(III) sulphate	$Fe_2(SO_4)_3$	brown
copper(II) sulphate	$CuSO_4$	blue
nickel chloride	$NiCl_2$	green

Figure 2
James Galway has a gold-plated flute. Perhaps it gives the music a mellow tone — but it also means that he doesn't have to polish it. Why not?

Key points

- Metals in the same group of the periodic table have very similar properties.
- Group I metals all float on water and react with it violently. They are all soft.
- Group II metals are harder and heavier than those in Group I. They are less reactive.
- Transition metals are heavy and generally have coloured compounds. Several of these metals and their compounds make good catalysts.

Quick questions

1 Predict what would happen if you held a sealed tube of caesium in your hand.

2 Describe a property that shows potassium is a more reactive metal than lithium.

★ **3** Describe what you think the reaction between rubidium and water might be like. Write the equation and name the products.

★ **4** Barium fizzes violently with dilute hydrochloric acid but very slowly with dilute sulphuric add. Why might this be?

5 How do you think you might tell sodium chloride and copper chloride apart, just by looking at them?

8.6 *Analysis of metals*

Atomic absorption spectrometry

Sometimes, scientists need to be able to detect very small amounts of metals that might not show up in ordinary tests. For example, the local water works might need to know whether river water contains lead, cadmium or mercury ions from local industrial wastes. These metals are very poisonous. Similarly, a public analyst or forensic scientist might want to detect minute amounts of poisonous metals in a dead animal or murder victim!

The instrument that they would use is the **atomic absorption spectrometer**. The sample under test is made into a solution and is then sprayed into a hydrogen flame. A beam of light from a lamp is shone through the flame and as the metal solution is vaporised, the metal atoms absorb some of the light from the lamp. (Figure 1.)

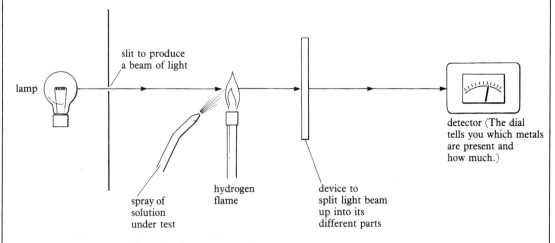

Figure 1 *Atomic absorption spectrometry.*

Each metal absorbs different parts of ordinary white light, so the remaining light coming out of the flame is examined by a detector. The detector is a 'black box' of electronics which can show which metal is in the solution, and even how much of it is there.

Flame tests

Some metal ions give distinct colours to a Bunsen flame when heated in the flame. The substance being tested must be moistened with concentrated hydrochloric acid. A small piece of platinum or nichrome wire is dipped into the moistened substance and then put into the side of the blue flame. (Figure 2.)

The colours are:

Li	crimson	Ca	brick red
Na	yellow	Sr	red
K	lilac	Ba	apple green
Rb	blue/red	Cu	blue/green
Cs	blue/red	Pb	grey/white

Figure 2
Doing a flame test. Why must the Bunsen flame be a hot blue flame?

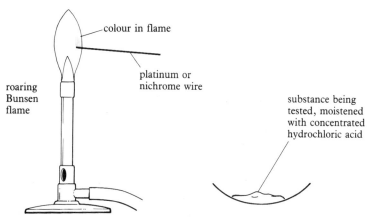

colour in flame

platinum or nichrome wire

roaring
Bunsen
flame

substance being tested, moistened with concentrated hydrochloric acid

Separating metal ions by ionic precipitation

If sodium hydroxide solution is added to a solution of a metal salt, the hydroxide ions will combine with the metal ions to form a **precipitate**.

Here is an example.

$$CuSO_4(aq) + 2NaOH(aq) \longrightarrow Cu(OH)_2(s) + Na_2SO_4(aq)$$
copper(II)   copper(II)
  sulphate   hydroxide

A precipitate of copper(II) hydroxide is formed. The reaction can be written more simply as an **ionic equation**.

$$Cu^{2+}(aq) + 2OH^-(aq) \longrightarrow Cu(OH)_2(s)$$

Distinguishing metal ions

Different metal ions react in different ways to sodium hydroxide. This sequence is used to distinguish between them.

• Put a small amount of the solution under test into a test tube.

- Add 2 drops of sodium hydroxide solution and look carefully at the contents of the tube.
- Now add an excess (three times the total volume of the test solution) of sodium hydroxide and stir thoroughly. Look at the contents again.

There are four things that could happen:

1 No precipitate at all is formed.
This means that the metal ions were Li^+, Na^+, K^+, Rb^+ or Cs^+. These metals ions do not form insoluble hydroxides. They can be distinguished from each other by flame tests.

2 A coloured precipitate is formed.
The ions could be:

     copper   blue
     iron (Fe^{2+}) pale green
     iron (Fe^{3+}) yellow/brown
     nickel   green
     chromium grey/green

3 A white precipitate is formed which dissolves again in excess sodium hydroxide solution. The ions could be:

     Al^{3+}, Zn^{2+} or Pb^{2+}

These can be distinguished by heating the precipitate in a test tube so that it dries out and changes into the oxide.

     e.g. $Zn(OH)(s) \longrightarrow ZnO(s) + H_2O(s)$
             zinc hydroxide   zinc oxide

     aluminium — oxide stays white
     lead   — oxide goes yellow
     zinc   — oxide goes yellow when hot but goes back to white when cold.

4 A white precipitate is formed which does not dissolve in excess sodium hydroxide solution. The ions could be

     Mg^{2+}, Ca^{2+}, Sr^{2+} or Ba^{2+}.

Calcium, strontium and barium can be distinguished from magnesium by their flame tests.

Key points

- Metal ions can be identified by ionic precipitation with sodium hydroxide solution.
- Some metals give distinctive flame colours when their compounds are heated in a flame.

Quick question

 * 1 Identify the follow metal ions from the description of what happened when the substance was mixed with sodium hydroxide solution.

(a) No precipitate formed, yellow flame.

(b) Blue precipitate, green flame colour.

(c) White precipitate, which is not soluble in excess sodium hydroxide, apple green flame.

(d) White precipitate, which dissolves in excess sodium hydroxide. Precipitate goes yellow when heated to dryness.

(e) Forms a yellow/brown precipitate.

(f) White insoluble precipitate, but no flame colour.

(g) Precipitate dissolves in excess sodium hydroxide. When heated, goes yellow and then white as it cools.

Questions

★ 1 Metals are extracted from their ores by reduction.

(a) Describe how
   (i) iron can be extracted from iron(III) oxide (Fe_2O_3) using a chemical reducing agent;
   (ii) aluminium can be extracted from aluminium oxide (Al_2O_3) using electrolysis.

(b) Why is aluminium **not** extracted using a chemical reducing agent?

(c) Both aluminium and iron are used to produce alloys.
   (i) What is an *alloy*?
   (ii) Name an alloy which is made from iron.
   (iii) Give two advantages of using an aluminium alloy for an aircraft body.

(d) Bauxite ($Al_2O_3.3H_2O$) is the main ore of aluminium.
   The relative atomic masses are Al 27, O 16, H 1.
   (i) Calculate the relative formula mass of bauxite.
   (ii) Calculate the percentage of aluminium by mass present in bauxite.
   (iii) Determine the maximum number of tonnes of aluminium that could be extracted from 5 tonnes of bauxite.   SEG

★ 2 A new metallic element segium has been discovered. Reactions are tried to find out how reactive segium is.

   Segium does **not** react with cold water.
   Segium, when heated, does react with steam.
   Aluminium will reduce segium oxide.
   Segium will **not** reduce magnesium oxide.
   Segium will **not** displace zinc from zinc sulphate solution.

(a) Which reaction shows that segium is
   (i) more reactive than copper;
   (ii) less reactive than potassium?

(b) Between which two metals should segium be placed in the Reactivity Series?   SEG

3 (a) The bar chart shows the percentage by mass of four of the more abundant elements in the Earth's crust.

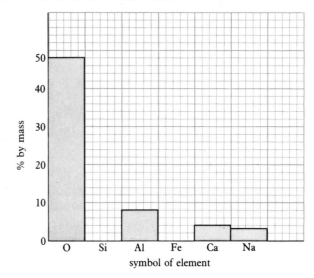

(i) Copy the bar chart and draw in silicon (27.6% by mass) and iron (5.1% by mass).

(ii) Use the bar chart to name the most abundant non-metal in the Earth's crust.

(iii) What is the percentage by mass of aluminium in the Earth's crust?

(b) Give **one** example where the use of aluminium saves energy. Explain why this use saves energy.

(c) Many aluminium products have the symbol below on the packaging.

(i) What is the meaning of the word *recyclable*?

(ii) Why is the recycling of aluminium so important?

(iii) Give **one** example of an aluminium article that can easily be collected for recycling.

(iv) How would you encourage large numbers of people to collect aluminium for recycling?

(d) Glass is another substance that can be recycled.

(i) Name another type of material that could be used in place of glass to make bottles.

(ii) Apart from cost give **one** advantage and **one** disadvantage of using your chosen material for bottles.

(iii) Apart from glass and aluminium name another substance that is recycled.

SEG

4 **(a)** What two substances must be present with iron for it to rust?

(b) How does a layer of tin prevent iron from rusting?

(c) Zinc blocks are attached to the steel hulls of ships, as shown in the diagram.

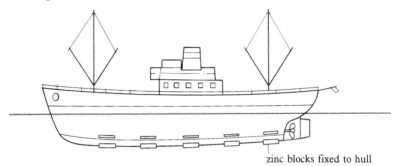

zinc blocks fixed to hull

    (i) How does the zinc protect the hull from rusting?
    (ii) What would be the effect of using copper blocks instead of zinc?

(d)   (i) What is an alloy?
    (ii) Give **one** example of an alloy and state what it is made of.
    (iii) Explain why the alloy you have named above is useful.

SEG

5 **(a)** From the reactivity series choose
    (i) an element that can occur uncombined in the earth's crust;
    (ii) an element that reacts vigorously with cold water.

(b) A gas may be made by passing steam over heated zinc using the apparatus shown below.

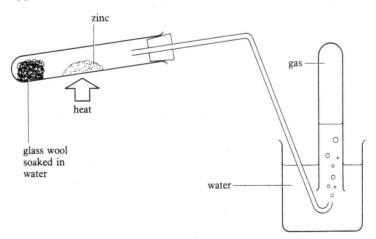

    (i) Name the gas collected.
    (ii) Name the other product of the reaction.
    (iii) Write an equation for the reaction.
    (iv) Name one metal that should **not** be reacted with steam in this way. Explain your choice.

SEG

6 (a) Give **one** reason why iron is the most widely used metal even though there is much more aluminium in the Earth's crust.

(b) A site has to be chosen to manufacture iron. The process needs
   1. large quantities of haematite (iron ore) which has to be imported;
   2. large quantities of coke;
   3. large quantities of limestone;
   4. a small amount of water.

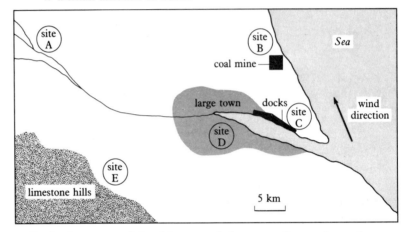

Study the map and the things needed to manufacture iron, then answer the questions.
   (i) For each of the sites A, B, C, D and E give **one** reason why it would **not** be a good choice.
   (ii) In spite of these disadvantages which site do you think would be best? Give your reasons.

(c) The main impurity in haematite (iron ore) is silicon dioxide. This is removed by reacting the ore with calcium oxide to form a molten slag (calcium silicate).
   (i) Which raw material is put into the blast furnace to form calcium oxide?
   (ii) Write an equation for the conversion of silicon dioxide into slag.

(d) Iron is produced in a blast furnace by reduction of iron(III) oxide.
   (i) Balance the equation for the reduction of iron(III) oxide by carbon monoxide.

   .... Fe_2O_3 + ... $CO \longrightarrow$ .... Fe + ... CO_2

   (ii) Why is this process called a *reduction*?

(e) A sample of molten iron (pig iron) from a blast furnace has the following composition.

Element	%	Element	%
sulphur	0.1	silicon	1
manganese	1	carbon	4
phosphorus	1	iron	

(i) Complete the table by writing in the percentage of iron.
(ii) Which physical property makes pig iron unsuitable for building a bridge?
(iii) When molten pig iron is coverted into steel which element is used to lower the percentage of carbon?

SEG

⋆ **7** The element rubidium is in Group I of the Periodic Table, immediately below potassium. Study the following scheme.

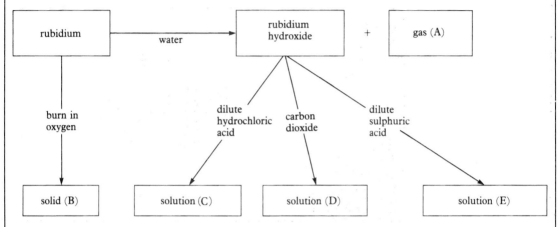

(a) Give the name or formula of
   (i) gas (**A**),
   (ii) solid (**B**),
   (iii) solution (**C**),
   (iv) solution (**D**),
   (v) solution (**E**).

(b) When dilute hydrochloric acid is added to solution (**D**), carbon dioxide is produced.
   Describe and state the result of a test to identify carbon dioxide.

(c) Which would react more vigorously with water, rubidium or potassium? Give a reason for your answer.

NEA

⋆ **8** (a) **A** $C + O_2 \longrightarrow CO_2$
        B $CO_2 + C \longrightarrow 2CO$
        C $Fe_2O_3 + 3CO \longrightarrow 2Fe + 3CO_2$
        D $CaCO_3 \longrightarrow CaO + CO_2$
        E $CaO + SiO_2 \longrightarrow CaSiO_3$

All the above reactions take place in the Blast Furnace. Write the letter of the correct answer to each of the following equations. Which equation

   (i) shows the formation of iron?
   (ii) involves the process of oxidation?
   (iii) shows the formation of slag?
   (iv) shows the reaction that is most likely to take place **first** in the Blast Furnace?

(b) Name the **three** materials put into the top of the Blast Furnace.

(c) Name **one** gas given out in the Blast Furnace.

(d) The corrosion of iron was investigated by giving six identical nails different treatments. One other nail was left untreated. All seven nails were then left for several weeks exposed to the atmosphere in a **small country town**. The results are given below.

Nail	Treatment	Cost of treatment	Mass of nail and coating BEFORE exposure to the atmosphere	Mass of nail and coating AFTER exposure to the atmosphere
A	Waxed	Low	5.0 g	5.3 g
B	Oiled	Low	5.0 g	4.1 g
C	Chromium plated	High	5.0 g	5.0 g
D	Painted	Low	5.0 g	5.4 g
E	Untreated		4.9 g	6.1 g
F	Galvanised	Fairly high	5.0 g	5.1 g
G	Dipped in salt solution	Low	5.0 g	6.7 g

which nail
(i) was **best protected** against corrosion?
(ii) received a treatment which made corrosion worse than it would have been had it been left untreated?
(iii) received a treatment which is usually used on iron railings?

(e) (i) In which case was there an obvious mistake in the weighing of the nail and the coating after the experiment?
(ii) Give **one** reason for your answer.

(f) Give **one** reason why the rate of corrosion of iron is greater in a heavily industrialised area.
Use the Periodic Table of the elements to answer parts **(g)** and **(h)**.

(g) How many of each of the following are there in one atom of iron?
(i) electrons
(ii) protons
(iii) neutrons

(h) Write the formula for each of the following compounds.
(i) iron(III) oxide
(ii) iron(III) fluoride

MEG

9 Organic chemistry

9.1 *Fossil fuels*

Why does gas smell?

The gas that comes out of Bunsen burners, gas fires and gas cookers needs to have a smell. If it didn't, you wouldn't be able to detect leaks, and that could be disastrous. Natural Gas, or North Sea Gas does not have any smell of its own, so the Gas Board put an artificial smell in. It's called Odorant BE and it consists of some very smelly, but harmless compounds. They are so smelly that very little is needed. As the gas enters the pipes, only 0.016 g of the compounds are added to each cubic metre of gas. Even this small amount is enough for most people to smell a gas leak.

Gas leaks can be very slow, but they can still be very dangerous. Sometimes the exact position of the leak is difficult to detect. For these conditions, the Gas Board use special electronic sniffers which detect the gas itself and not the smelly compounds. They are a thousand times more sensitive than the human nose and are usually used at night to detect gas leaking up through pavements and roads.

- Why are electronic sniffers used at night?
- Why doesn't the Gas Board give gas a pleasant smell like roses?

What to do if you smell gas

1 Do not use matches or a naked flame to find the leak.
2 Do not operate electrical switches — either on or off.
3 Open doors and windows, to get rid of the gas.
4 Check to see if a tap has been left on accidentally, or if a pilot light has gone out. If not, there has probably been a gas leak. So turn off the whole supply at the meter and call the gas service.
5 If you can't turn off the supply, or the smell continues after you have, you **must** call the gas service, or ask someone else to help you do so.

Figure 1
A gas sniffer van detecting escaping gas.

Why 'fossil' fuels?

Coal, oil and natural gas are called **fossil fuels** because they were made from living things millions of years ago, and have been preserved under the ground.

Coal was formed over 200 million years ago from giant forests. As plants and trees died and fell to the ground, they were covered with silt and mud from rivers and floods, and slowly pressed down into the ground. Changes in the Earth's structure compressed the dead plant matter and heated it until it slowly changed into the black substance we know as coal.

Coal is now found in layers or seams under the ground. Coal of different ages has different compositions. The table below lists the types of coal found in this country.

Type of coal	Content percentage (%)		
	carbon	hydrogen	oxygen
lignite (brown coal from Devon)	67	5	28
bitumous (from England and Scotland)	88	6	6
anthracite (hard black coal from South Wales)	94	3	3

Oil and **gas** are younger fossil fuels. 50 million years ago, most of the surface of the Earth was covered with water and the seas were full of tiny marine animals and plants called plankton. As they died, their bodies fell to the bed of the oceans. Over hundreds of thousands of years they were covered with silt and sediments and were slowly heated and compressed just like coal. As they slowly changed into oil, the liquid soaked into porous rocks (these soak up liquids) which were layered between non-porous rocks so that the rock oil was trapped. (Figure 2.) Gases, mainly methane, were formed too but dissolved in the oil under the high pressure.

Figure 2
Oil and gas are found in this type of geological structure.

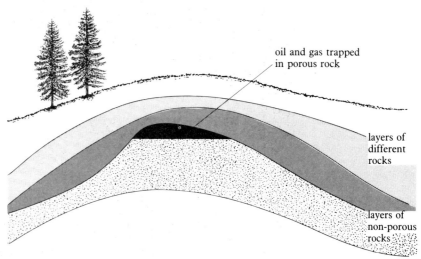

oil and gas trapped in porous rock

layers of different rocks

layers of non-porous rocks

Mining coal

Most coal in Britain is found deep in the ground and must be mined. Shafts are sunk into the ground (the average depth is about 450 m, although the deepest one is 1000 m) and everything — men, machines and coal, is taken up and down in lifts called cages.

Not all coal is mined in this way. Some coal seams are near the surface and can be worked by open cast mining. Coal scientists are researching into ways of dissolving the valuable chemicals out of the coal without digging it up, or even burning it underground as a fuel and just bringing the heat to the surface.

Drilling for oil

In 1985, 58 million barrels of oil a day (a barrel would fill four car petrol tanks), were produced throughout the World.

Drilling is a very expensive business and so geologists have to make sure that there is oil there before drilling begins. First of all, possible areas are surveyed from the air. Special geological features can be spotted more easily this way. Then a **seismic survey** is carried out. (Figure 3.) A small explosion is set off in the ground and the vibrations from the different layers of rock under the ground are detected and analysed. Next, if the area still seems suitable, geologists make a **test bore** to examine samples of rock to look for traces of oil or fossils.

Figure 3
The vibrations from the explosion take different lengths of time to rebound from the layers of rocks. This process is called a seismic survey.

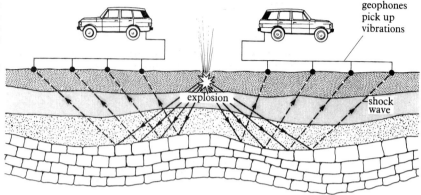

Finally, if oil seems a possibility, a **derrick** is set up. (Figure 4.) The tall structure supports a pulley so that drill pipes, each 10 m in length, can be connected to the drill bit as it is driven down into the ground. More and more pipes are added as the drill goes deeper, until the drill bit has to be changed. Then everything has to be brought to the surface, piece by piece, and the whole process started again! All this is much more dangerous in the North Sea where there are frequent gales and the oil rig may be standing on huge legs deep in the water, or even floating on the surface.

When oil is finally found, it may be under great pressure, so a valve must be fitted to the top of the hole. The oil and gas can be safely pumped up and taken to **oil refineries** or piped to gas terminals ashore.

Figure 4
An oil derrick. A tall structure is needed to fit the drill pipes together as the drill bit cuts deeper into the ground.

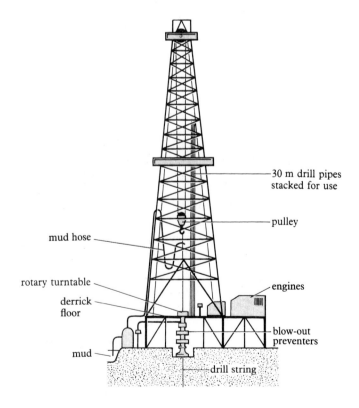

Refining oil

When it comes out of the ground, **crude oil** is a mixture of many different chemicals which are mainly **hydrocarbons**. Hydrocarbons are made of chains of carbon atoms, with hydrogen atoms attached to them. (You can read more about hydrocarbons in the next section.) They differ in the size of their molecules, and in the number of carbon atoms in their chains.

Hydrocarbons with short chains, only a few carbon atoms in them, have low boiling points. They are liquids which boil very easily, or gases. Hydrocarbons with long chains and many carbon atoms are thick and oily and have high boiling points. They can be separated by **fractional distillation** where each liquid is boiled out at its own boiling point.

Fractional distillation

This is done in a **fractionating column**. (Figure 5.) Crude oil is heated in a furnace to 800°C. Its vapours are then fed into the column. The most **volatile** hydrocarbons (those that boil most easily, with low boiling points) find their way to the top. The least volatile liquids do not get as high, and some sink to the bottom. In the column are trays in which the vapours condense to liquids. The liquids in each tray have the same range of boiling points. The different liquids are called **fractions**. Each tray has holes covered with small domes called **bubble caps**. This enables vapours to get through the condensed liquids. The different liquids sort themselves out because the vapours going up try to boil the liquids in the trays, and the liquids trickling down try to condense the vapours. This is how fractional distillation works.

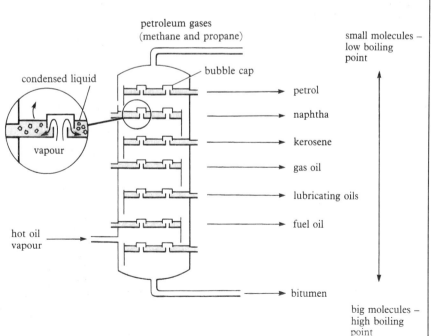

Figure 5
This fractionating column separates liquids of different boiling point. The liquids which boil first at the lowest temperatures get furthest up the tower.

petroleum gases
(methane and propane)

condensed liquid

bubble cap

vapour

hot oil vapour

petrol

naphtha

kerosene

gas oil

lubricating oils

fuel oil

bitumen

small molecules – low boiling point

big molecules – high boiling point

The fractions from the trays are removed and taken away for further treatment. The table below shows some of the things that are made from crude oil.

Figure 6
*The fractions obtained
from the fractional
distillation of oil.*

Fraction	Boiling point range	Number of carbon atoms in chain	Some uses
refinery gas dissolved in the oil	less than 40°C	1–4	fuels e.g. calor gas camping Gaz
petrol and naphtha	40°C–170°C	4–12	fuel for cars, chemicals for plastics and insecticides
kerosene	150°C–240°C	9–16	paraffin, jet fuel
diesel oil	220°C–250°C	15–25	fuel for lorries
lubricating oils	250°C–350°C	20–70	polishes and waxes, oil for machinery
fuel oil	above 350°C	around 60	fuel for ships and power stations
bitumen	solid	around 60	tar for roads, weather-proofing materials

Using fossil fuels

More than three quarters of the coal mined or imported into the UK is used in power stations to generate electricity. When coal is burned, the heat is used to make steam. The steam powers turbines and generators. Unfortunately, the steam is still very hot after it has been used and this heat is wasted. The gigantic cooling towers seen in Figure 8 are there to condense the steam so that it does not escape into the air.

The rest of the coal is used as domestic fuel, for industries, or is exported. Some coal is heated in big retorts without being allowed to burn so that all the valuable chemicals in it are driven off and condensed. The **coke** that is left is used as smokeless fuel and is used for making iron in blast furnaces.

What makes a good fuel?

Fuels are burned to produce heat. The heat is used directly as in a coal fire, or used to heat up water as in a gas or oil central heating system. Fuel is also used to make steam for power stations. Sometimes the heat produced by fuel is used in an engine to produce movement, as in cars and trains. In each case, the fuel needs to be suitable. Here are some guidelines for deciding if a fuel is suitable.

- How much heat does it produce?
- Is it cheap?

- Is it easy to obtain?
- Is there much of it?
- How easy is it to transport and store the fuel?
- Does it produce dangerous fumes or smoke?
- Does it leave behind a lot of ash?

Figure 7
The DRAX power station in North Yorkshire will use 37 000 tonnes of pulverised coal each day and generate 4000 MW of electricity. How many homes could that supply?

Figure 8
Waste steam from the power station is condensed in these cooling towers and the water is re-used or put in the river.

Alternatives to fossil fuels

When fossil fuels eventually run out, alternative sources of energy will be needed. Here are some of them.

Solar energy can be used to heat water in solar panels on roofs to supply household hot water. Even if the sun doesn't shine, energy from the sun gets through the clouds. Alternatively, the sun's light can be converted to electricity with solar cells.

Wind energy can be converted into electricity by giant propellors. Banks of wind generators like the ones in the photograph, placed in windy places such as Orkney and other Scottish Western islands could produce a lot of electricity.

Figure 9
This turbine has blades 825 metres long! It produces enough electricity to light 13 thousand light bulbs.

Waves and tides have energy that could be used. In places where tides are high, the rising and falling water could power generators. Plans are being made to build a barrage across the mouth of the river Severn. Huge tides here could generate up to 10% of Britain's electricity and bring much needed industry to the area. However, not only would it cost huge sums of money to build, but environmentalists fear that damage might be caused to the river.

Biological sources such as **Biogas**, a mixture of methane and other flammable gases, made from animal manure by the action of bacteria. Ethanol, produced by the fermentation of sugar is used instead of petrol in Brazil.

Key points

- Coal, oil and gas are fossil fuels.
- Coal and oil are expensive to extract from the ground.
- Oil is a mixture of many chemicals and must be separated by fractional distillation.
- There are many alternative sources of energy besides fossil fuels.

Quick questions

1 The Drax power station uses coal which has been pulverised instead of in lumps. Why is this?

2 Oil drilled in the north of Canada (where the temperature is often below freezing point) is transported in pipes which are electrically heated. Why do you think this is necessary?

3 Hydrogen, coal and candle wax are all fuels. Describe some good things and bad things about each of them.

9.2 *Hydrocarbons*

Lead free petrol

In a car engine, petrol vapour is mixed with air and, after being compressed, is ignited with a spark from the spark plug. The large volume of gases formed pushes the piston down in the cylinder. The piston is connected to a crank, and this, via the crank-shaft and gear-box, makes the wheels go round.

Petrol is a mixture of chemicals called **hydrocarbons**. This mixture contains highly volatile liquids that will vapourise and ignite at low temperature so that cars will start and run on cold days. It must also contain liquids which are less volatile so the fuel does not boil in the cylinder and separate before it has a chance to burn.

Knocking, or pinking as it is sometimes called, occurs when the petrol explodes in the cylinder rather than burning smoothly. This gives the piston a kick rather than a steady push and can cause damage to the big end of the engine. The octane rating of a fuel is a measure of its resistance to knocking.

Modern cars have a high compression ratio (this is a measure of how much the petrol/air mixture is compressed compared with the distance the piston travels). High compression ratio engines need a high octane fuel. The octane rating of petrol has been increased by adding compounds such as tetramethyl lead and tetraethyl lead. The big disadvantage of this, though, is that most of the lead comes out of the exhaust pipe as lead dust and causes pollution. If the lead is removed from petrol, then changes are needed in the design of engines. The adapted engines would use lower octane petrol or petrol in which the anti-knock compounds do not contain lead. These compounds are more expensive and so the fuel will cost more.

Figure 1
Cars pollute the atmosphere with lead from their exhaust fumes. However, banning lead in petrol is not as simple a matter as it might seem.

Alkanes

Organic chemicals belong to families called **homologous series**. Homologous means 'same structure' and the members of the family of alkanes all have carbon atoms with four covalent bonds. The table below gives information about the homologous series called **the alkanes**.

Name	Number of carbon atoms	Formula	Structure
methane	1	CH_4	$H-\overset{\displaystyle H}{\underset{\displaystyle H}{C}}-H$
ethane	2	C_2H_6	$H-\overset{\displaystyle H}{\underset{\displaystyle H}{C}}-\overset{\displaystyle H}{\underset{\displaystyle H}{C}}-H$
propane	3	C_3H_8	$H-\overset{\displaystyle H}{\underset{\displaystyle H}{C}}-\overset{\displaystyle H}{\underset{\displaystyle H}{C}}-\overset{\displaystyle H}{\underset{\displaystyle H}{C}}-H$
butane	4	C_4H_{10}	$H-\overset{\displaystyle H}{\underset{\displaystyle H}{C}}-\overset{\displaystyle H}{\underset{\displaystyle H}{C}}-\overset{\displaystyle H}{\underset{\displaystyle H}{C}}-\overset{\displaystyle H}{\underset{\displaystyle H}{C}}-H$
pentane hexane heptane octane . . .			

You can work out the formula of any member of the family by using the general formula C_nH_{2n+2}, where n is the number of carbon atoms. As you proceed through the series, the molecules get bigger, and so their boiling points increase. At room temperature hexane is a liquid and $C_{20}H_{42}$ is a solid. (See Figure 2.)

Figure 2
The melting and boiling points of alkanes increase with the size of their molecules.

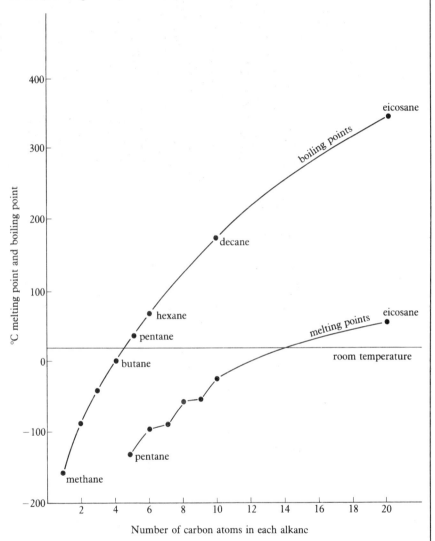

Diagrams of alkanes in books are always drawn flat, but in fact, methane has a tetrahedral shape (look at Figure 3) and the other alkanes have similar structures.

Figure 3
The methane molecule has a tetrahedral shape, but this is difficult to draw. The structure on the right shows how methane is usually drawn.

Isomerism

The hydrocarbon C_4H_{10} has two structures. (See Figure 4.) The different structures are called isomers.

Figure 4
These molecules are isomers. They have the same formula but different structures.

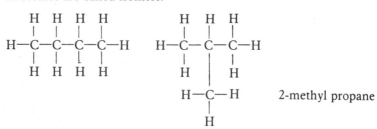

2-methyl propane

> Isomers are compounds having the same formula but different structures.

Uses of the alkanes

All hydrocarbons burn oxygen and the first four members of the alkanes are all used as fuels.

Methane is the main constituent of natural gas. **Propane** is liquefied by gentle pressure and used instead of methane in some areas, or to power specially converted cars.

Butane can also be liquefied and sold as **Calor Gas** and **Camping Gaz**.

In a good supply of air, these hydrocarbons burn to form only carbon dioxide and steam.

$$CH_4(g) \; + \; 2O_2(g) \longrightarrow \; CO_2(g) + \; 2H_2O(g)$$
methane

$$2C_4H_{10}(g) + 13O_2(g) \longrightarrow 8CO_2(g) + 10H_2O(g)$$
butane

If they are burnt in a restricted supply of air, some of the carbon many remain as soot, or more dangerously, only be oxidised to **carbon monoxide**, a very poisonous gas.

The balanced flue

Gas fires which are fitted into rooms in which there is no chimney have to be of a special sort. They have an outlet through the wall so that all the air needed for burning comes in from the outside, and all the combustion fumes go outside. Only the heat comes inside the room. In this way, there is no chance of carbon monoxide poisoning. (Figure 5.)

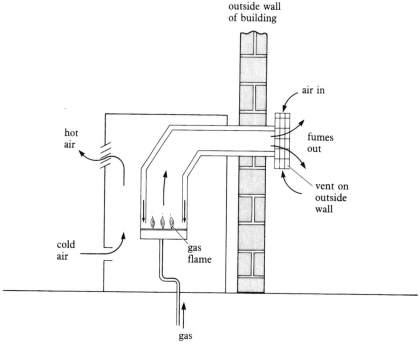

Figure 5
*A balanced gas flue.
None of the gases from
the flame can get into the
room. Why is this safety
device necessary?*

Alkenes

Not all hydrocarbons have single bonds between their carbon atoms. One group, called alkenes has double bonds.

Ethene is a member of the family of alkenes. It has a double bond.

$$C_2H_4 \qquad \begin{array}{c} H \\ \diagdown \\ H \diagup \end{array} C = C \begin{array}{c} H \\ \diagup \\ \diagdown H \end{array}$$

The laboratory preparation of ethene

Figure 6 shows the apparatus required

1 Cotton wool is soaked in ethanol and placed at the bottom of a test tube.

2 Some aluminium oxide is placed half way up the test tube. It is to act as a catalyst.

3 The cotton wool is only warmed but the catalyst must be heated. Ethanol vapour passes over the aluminium oxide and is split into ethene and steam.

$$\underset{\text{ethanol}}{C_2H_5OH(g)} \longrightarrow \underset{\text{ethene}}{C_2H_4(g)} + H_2O(g)$$

4 These gases are bubbled through water where the steam condenses and the ethene can be collected.

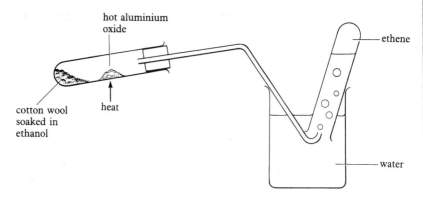

Figure 6
*The dehydration of
ethanol to make ethene.
What happens to any
ethanol vapour that is
not dehydrated?*

The commercial production of ethene

Ethene is made by breaking down **naphtha** from crude oil in a process called **cracking**. Naphtha contains fairly big molecules with carbon chains up to 10 atoms long. The naphtha is mixed with steam and heated in a furnace to a temperature of 800 °C, sometimes over a catalyst of nickel. The naphtha breaks down into gasoline and ethene, propene, hydrogen, and other gases. These are separated by condensing and then fractionally distilling them.

Substitution and addition reactions

Alkanes

Alkanes are said to be **saturated hydrocarbons** because they do not contain any double bonds. When these compounds react, their hydrogen atoms have to be replaced by other atoms.

When methane is mixed with chlorine gas in bright light, **chloromethane** is slowly formed.

$$\underset{\text{methane}}{H-\overset{\displaystyle H}{\underset{\displaystyle H}{C}}-H} + Cl_2 \xrightarrow{\text{light}} \underset{\text{chloromethane}}{H-\overset{\displaystyle H}{\underset{\displaystyle H}{C}}-Cl} + \underset{\text{hydrogen chloride gas}}{HCl}$$

If ethane gas is bubbled through liquid bromine over a catalyst of iron, **bromoethane** is slowly formed and hydrogen bromide gas is released.

$$\underset{\text{ethane}}{H-\overset{\displaystyle H}{\underset{\displaystyle H}{C}}-\overset{\displaystyle H}{\underset{\displaystyle H}{C}}-H} + Br_2 \xrightarrow[\text{catalyst}]{\text{iron}} \underset{\text{bromoethane}}{H-\overset{\displaystyle H}{\underset{\displaystyle H}{C}}-\overset{\displaystyle H}{\underset{\displaystyle H}{C}}-Br} + \underset{\substack{\text{hydrogen} \\ \text{bromide gas}}}{HBr}$$

Both these reactions are called **substitution reactions**.

> Saturated hydrocarbons undergo substitution reactions.

Alkenes

Ethene is an **unsaturated hydrocarbon** because there is a **double bond** between its carbon atoms. When it reacts, the double bond splits into two single bonds.

When ethene gas is bubbled through a yellow solution of bromine in water, the solution goes colourless and dibromoethane is formed.

$$
\underset{\text{ethene}}{\begin{array}{c} H \\ H \end{array}\!\!C\!=\!C\!\!\begin{array}{c} H \\ H \end{array}} + Br_2(aq) \longrightarrow \underset{\text{dibromoethane}}{Br\!-\!\overset{\displaystyle H}{\underset{\displaystyle H}{C}}\!-\!\overset{\displaystyle H}{\underset{\displaystyle H}{C}}\!-\!Br}
$$

This is called an **addition reaction**.

> Unsaturated hydrocarbons undergo addition reactions.

If ethene is mixed with hydrogen gas and is heated over a catalyst of nickel or platinum, ethane is formed.

$$
\underset{\text{ethene}}{\begin{array}{c} H \\ H \end{array}\!\!C\!=\!C\!\!\begin{array}{c} H \\ H \end{array}} + H_2 \xrightarrow[\text{Ni}]{\text{catalyst}} \underset{\text{ethane}}{H\!-\!\overset{\displaystyle H}{\underset{\displaystyle H}{C}}\!-\!\overset{\displaystyle H}{\underset{\displaystyle H}{C}}\!-\!H}
$$

This addition reaction is also called **hydrogenation**.

Polymerisation

If ethene gas is heated under pressure, it changes from an invisible gas to a waxy white solid.

$$
\underset{}{\begin{array}{c} H \\ H \end{array}\!\!C\!=\!C\!\!\begin{array}{c} H \\ H \end{array}} \longrightarrow \cdots C\!-\!C\!-\!C\!-\!C\!-\!C\!-\!C \cdots
$$

The double bonds in the ethene have been opened up and the thousands of ethene molecules have joined together to make a very long chain alkane.

This process is called **polymerisation**. Ethene is called a **monomer**, and the product is called a **polymer**. In this reaction, the polymer is called **poly(ethene)**.

$$\text{ethene} \longrightarrow \text{poly(ethene)}$$
$$\text{(a monomer)} \qquad \text{(a polymer)}$$

Many other polymer substances can be made in the same way, starting from monomers with double bonds in their molecules.

chloroethene turns into **poly(chloroethene)**, better known as **PVC**. (Its old name was polyvinyl chloride.)

$$
\begin{array}{c}
\text{H} \diagdown \qquad \diagup \text{Cl} \\
\text{C}=\text{C} \\
\text{H} \diagup \qquad \diagdown \text{H}
\end{array}
\longrightarrow
\cdots
\begin{array}{cccccc}
\text{H} & \text{Cl} & \text{H} & \text{Cl} & \text{H} & \text{Cl} \\
| & | & | & | & | & | \\
\text{C}-&\text{C}-&\text{C}-&\text{C}-&\text{C}-&\text{C} \\
| & | & | & | & | & | \\
\text{H} & \text{H} & \text{H} & \text{H} & \text{H} & \text{H}
\end{array}
\cdots
$$

phenylethene turns into **poly(phenylethene)**, which is more usually called by its old name of **polystyrene**.

$$
\begin{array}{c}
\text{H} \diagdown \qquad \diagup \text{C}_6\text{H}_5 \\
\text{C}=\text{C} \\
\text{H} \diagup \qquad \diagdown \text{H}
\end{array}
\longrightarrow
\cdots
\begin{array}{cccccc}
\text{H} & \text{C}_6\text{H}_5 & \text{H} & \text{C}_6\text{H}_5 & \text{H} & \text{C}_6\text{H}_5 \\
| & | & | & | & | & | \\
\text{C}-&\text{C}-&\text{C}-&\text{C}-&\text{C}-&\text{C} \\
| & | & | & | & | & | \\
\text{H} & \text{H} & \text{H} & \text{H} & \text{H} & \text{H}
\end{array}
\cdots
$$

Polystyrene is a brittle, glasslike plastic. If air is blown into it whilst it is being made, **expanded polystyrene** is formed. This is the light-weight plastic used for packaging or for heat insulation.

Each of these polymers has been made by **addition polymerisation** because a double bond in the monomer has been made into single bonds.

Nylon (poly(amide)) and **Terylene** (poly(ester)), are made by a slightly different process called **condensation**. Nylon is made when hexane-1,6-diamine solution is added to hexanedioic acid solution.

$$
\text{NH}_2-
\begin{array}{cccccc}
\text{H} & \text{H} & \text{H} & \text{H} & \text{H} & \text{H} \\
| & | & | & | & | & | \\
\text{C}-&\text{C}-&\text{C}-&\text{C}-&\text{C}-&\text{C} \\
| & | & | & | & | & | \\
\text{H} & \text{H} & \text{H} & \text{H} & \text{H} & \text{H}
\end{array}
-\text{NH}_2 \; + \;
\begin{array}{c}
\text{O} \\
\diagdown \\
\text{C}-\text{C}-\text{C}-\text{C}-\text{C}-\text{C} \\
\diagup \\
\text{HO}
\end{array}
\begin{array}{c}
\diagup \text{O} \\[-2pt] \\
\diagdown \text{OH}
\end{array}
$$

$$
\cdots -\text{N}-
\begin{array}{cccccc}
\text{H} & \text{H} & \text{H} & \text{H} & \text{H} & \text{H} \\
| & | & | & | & | & | \\
\text{C}-&\text{C}-&\text{C}-&\text{C}-&\text{C}-&\text{C} \\
| & | & | & | & | & | \\
\text{H} & \text{H} & \text{H} & \text{H} & \text{H} & \text{H}
\end{array}
-\text{N}-
\begin{array}{cccccc}
\text{O} & \text{H} & \text{H} & \text{H} & \text{H} & \text{O} \\
\| & | & | & | & | & \| \\
\text{C}-&\text{C}-&\text{C}-&\text{C}-&\text{C}-&\text{C} \\
& | & | & | & & \\
& \text{H} & \text{H} & \text{H} & &
\end{array}
- \cdots
$$

There are several different forms of nylon produced from slightly different monomers.

Why 'plastics'?

There are two kinds of plastics, **thermoplastics** and **thermosetting** plastics.

Poly(ethene), PVC, polystyrene and Nylon are all hard although they may be bendable substances. They are pliable and mouldable, and that is what 'plastic' means. They are thermoplastics, because when they are heated, they become soft, and when cooled, they become hard again. This happens because their long polymer chains are intertwined with each other. When heated, the long chains move about with the added energy and slide over each other, allowing the plastic to become mouldable. Thermoplastics do not have a fixed melting point, but a softening range instead.

Thermosetting plastics set hard once they have been heated during their manufacture. They do not soften again, no matter how much they are heated. This is because, during their manufacture, linkages between their polymer chains are formed, stopping the chains moving so easily. Thermosetting plastics are suitable for things that get hot during their use.

Uses of thermosetting and thermoplastics

Plastic	Uses
thermoplastics:	
poly(ethene), or polythene	squeezy detergent bottles, rubbish bags, milk bottle crates, freezer bags
poly(propene), or polypropylene	film for packaging, snap-on lids, bottles
PVC	drain pipes, insulation for electric wiring, imitation leather furniture, water-proof clothes, shoes
polystyrene	yoghurt cartons, disposable cups
expanded polystyrene	ceiling tiles, packing material, insulation for fridges and coolboxes
acrylic plastics	safety glass, traffic signs, car reflectors and indicators
polycarbonate	babies' bottles
poly(tetrafluoroethene) called PTFE	non-stick surfaces
poly(amide), called Nylon	curtain rails, hinges, fibres for clothes, rope
poly(ester), called Terylene	fibres for clothes, video tape, plastic bottles
polyurethane foam	stuffing for furniture
thermosetting plastics:	
urea methanal	electric light fittings
melamine methanal, called Melamine	tableware
phenol methanal, called Phenolic	Formica, saucepan handles

Shaping plastics

Thermoplastics can be squeezed, stretched or squashed into shape when they are hot. Many thermoplastics are shaped by **extrusion**. Small granules of the plastic are heated and forced through a nozzle. The plastic comes out in a continuous length, the same shape all the way through.

Nylon and Terylene fibres, curtain railing, drain pipes, outer casing for electric wires, sheets of acrylic for safety windows, polythene tubing, are all made in this way.

Milk bottle crates, buckets, telephones, steering wheels, soap dishes and safety helmets are all made by **injection moulding**. Hot plastic is forced into a mould. In **blow moulding**, the plastic is forced into the mould by compressed air. Plastic bottles are made in this way. (Figure 7.) In **vacuum moulding**, the plastic is sucked into the mould and a lot of packaging uses this technique. Plastic film for food bags, raincoats, wallpaper, furniture is made by **calendering** where the plastic is heated and squeezed between a series of rollers until it is the right thickness.

Thermosetting plastics have to be made into the right shape as they are manufactured, because once made, their shape cannot be changed by heating. Formica work tops are made by **laminating** in which layers of paper, plastic and cardboard are sandwiched together and squashed in a giant press. Melamine cups are made by **compression moulding** in which powdered chemicals are heated as they are compressed in a mould of the correct shape.

Figure 7
Thermoplastics can be blown into shape when they are hot and soft.

Figure 8
*Squash can now be a
spectator sport. This
court is made of Perspex.
Do you think it shatters
easily?*

Recycling plastics

Plastics have many advantages over other materials like stone, metal and wood. They are often cheaper, more flexible, lighter, can be coloured more easily, do not corrode, are water proof and do not conduct electricity. But they do have a big disadvantage, they do not rot. So, when plastics are taken to the rubbish tip along with food, paper and other household rubbish, they stay in the ground and do not decay away.

There are ways around this problem. For example they could be separated from the rubbish before it is buried. Reclaimed thermoplastics can be used to make such things as black plastic sheeting for dustbin bags.

Another way of disposing of plastics is to burn them along with other rubbish. About 7% of the waste plastic in this country is burned, and the heat is used for industrial heating or in a power station to make electricity. However, burning plastics has its problems. If the temperature is not high enough, poisonous and acidic gases may be given off as well as a lot of smoke.

Key points

- Hydrocarbons are compounds containing hydrogen and carbon.
- Alkenes are hydrocarbons that contain double bonds.
- Ethene may be made by dehydrating ethanol. Industrially, it is made by cracking naphtha.
- Alkanes are saturated hydrocarbons; alkenes are unsaturated hydrocarbons.
- Alkanes undergo substitution reactions; alkenes undergo addition reactions.

- Plastics can be thermosetting or thermoplastic.
- Polymers are long chain molecules made from small monomer molecules.
- Thermoplastics may be re-shaped when hot. Thermosetting plastics are shaped as they are made and cannot be re-shaped by heating.

Quick questions

1 Write the formula and draw the structure of an alkane with eight carbon atoms. What is its name?

2 Predict what compound might be made if ethene is mixed with chlorine gas. Draw the structure of the new compound.

3 Write the equation for the combustion of ethane. Why would it be dangerous if you did not allow enough air to get into the reaction?

4 What plastics would be suitable for making (a) table mats (b) shampoo bottles (c) space helmets. What method would you use to manufacture each of these items?

5 Why do people so often die in house fires, even though they might not be anywhere near the flames?

9.3 *Ethanol and ethanoic acid*

Chips without vinegar?

We are used to fresh fish, but before the days of freezers, fish would sometimes start to go bad before it was eaten. The first sign that fish is starting to decompose is the strong 'fishy' smell when compounds called **amines** are formed. They smell strongly of ammonia too and like ammonia, they are alkaline.

Vinegar is a solution of **ethanoic acid**, and acids neutralise alkalis to form salts. The effect of putting ethanoic acid onto alkaline amines would be to neutralise the amines and remove their smell. This would disguise the fact that the fish was going bad. One household tip long ago was to wash fish in vinegar. Of course, we like the taste of vinegar too — what would chips taste like without it?

Ethanol

Ethanol is a member of the family of **alcohols**. Like the alkanes, the simplest member of the group has a molecule that contains only one carbon atom.

methanol	CH_3OH	$H-\overset{\displaystyle H}{\underset{\displaystyle H}{C}}-OH$
ethanol	C_2H_5OH	$H-\overset{\displaystyle H}{\underset{\displaystyle H}{C}}-\overset{\displaystyle H}{\underset{\displaystyle H}{C}}-OH$
propanol	C_3H_7OH	$H-\overset{\displaystyle H}{\underset{\displaystyle H}{C}}-\overset{\displaystyle H}{\underset{\displaystyle H}{C}}-\overset{\displaystyle H}{\underset{\displaystyle H}{C}}-OH$

and so on.

Methanol is a poisonous compound made industrially from methane. Ethanol is made industrially by the reaction of ethene with steam under pressure with a catalyst of sulphuric acid.

$$\underset{\text{ethene}}{\overset{H}{\underset{H}{\Large{>}}}C=C\overset{H}{\underset{H}{\Large{<}}} + H_2O \longrightarrow \underset{\text{ethanol}}{C_2H_5OH}$$

It is also made naturally by the process of **fermentation**.

Fermentation

Sugars, like sucrose and glucose, are broken down in solution by the action of yeast cells. Biological catalysts called **enzymes** found in the yeast control the reactions. In fermentation glucose is converted into ethanol, and carbon dioxide gas is released.

$$\underset{\text{glucose}}{C_6H_{12}O_6} \longrightarrow \underset{\text{ethanol}}{2C_2H_5OH} + 2CO_2(g)$$

Amateur winemakers use the apparatus shown in Figure 1. The jar contains fruit extract (grapes in real wine), water and yeast. Sometimes extra sugar is added too. The mixture is kept warm, to help the yeast grow, but not too hot or the yeast will die. As the reaction takes place, carbon dioxide is released through the air lock. Air cannot get in. If it did, bacteria might grow and the wine would oxidise to vinegar. Sometimes a **yeast nutrient** is added to help the yeast grow. This contains a chemical similar to that in fertilizers such as ammonium phosphate.

When the ethanol reaches a particular concentration, it kills the yeast and the fermentation stops. There is always a thick sediment of dead yeast cells at the bottom of the jar.

When **beer** is made, malt is fermented. **Malt** is extracted from barley seeds that have just sprouted and then have been roasted. The mixture is fermented with yeast, just like wine, in big vats at a carefully controlled temperature. (Figure 2.) The barley that is left over is often sold as cattle food and the left over yeast is used to make Marmite.

The ethanol obtained by fermentation is a dilute solution in water. Much more concentrated solutions can be made by **fractional distillation**.

Figure 1
Amateur winemakers use fermentation jars like this. The air lock on top keeps air out, and the wine fresh. What chemical reaction will take place if air does get in for a long time?

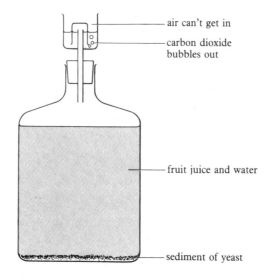

air can't get in

carbon dioxide bubbles out

fruit juice and water

sediment of yeast

Figure 2
Beer is fermented malt. Hops are added to give flavour and in earlier times, to stop the beer going bad.

Fractional distillation

The liquid made by fermentation is a mixture of ethanol and water. These two liquids have different boiling points:

ethanol boils at   78°C
water boils at   100°C

Look at Figure 3. When the water and ethanol mixture is boiled a mixture of ethanol vapour and steam is formed.

- Which is the more volatile compound?
- Which vapour will get to the top of the fractionating column most easily without condensing?
- Which vapour will condense most easily and drip back into the flask?

Figure 3
The liquid in the beaker is called the distillate. What does it consist of? What temperature do you think will show on the thermometer?

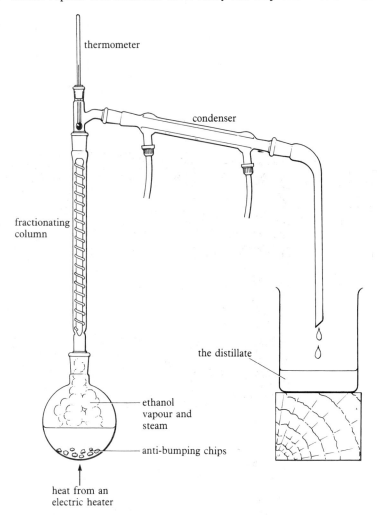

thermometer

condenser

fractionating column

the distillate

ethanol vapour and steam

anti-bumping chips

heat from an electric heater

Drinks like whisky, gin, rum and vodka are made by fractional distillation. Whisky is fermented malt that has been distilled in copper stills like those in Figure 4. Besides occuring naturally in alcoholic drinks, ethanol is important as an antiseptic, as a solvent, and as a fuel.

Figure 4
Whisky is made by the fractional distillation of fermented malt in copper stills.

Methylated spirits is industrially produced ethanol that has a small amount of methanol in it. It also contains a colouring agent, so that you don't drink it by mistake. It burns with a clean flame and is a very useful fuel.

$$C_2H_5OH + 2O_2(g) \longrightarrow 2CO_2(g) + 3H_2O(g)$$

Alcohol abuse

The existence and effect of alcohol (ethanol) have been known for thousands of years. Lots of people drink some form of alcoholic drink from time to time. But, drinking alcohol has some side-effects. After only one pint of beer or a couple of glasses of wine, reactions are slowed down. Judgement of distance and speed become less accurate and after drinking alcohol people are more likely to have accidents with machinery and cars.

In the UK it is against the law to drive a car or motor bike if the alcohol level in your blood is greater than 80 milligrams of alcohol per 100 cm^3 of blood. The penalty for this is disqualification from driving for one year and up to £1000 fine, or six months in prison.

When someone drinks alcohol, the body breaks it down slowly and the side-effects wear off. If someone consumes a lot of alcohol, there is too much for the body to cope with and the person can be very ill — even die. Alcohol shrinks tissue in your brain and damages the liver. Heavy drinkers often get a disease called **cirrhosis** of the liver.

Ethanoic acid

Ethanoic acid is made when ethanol is oxidised. Sometimes this happens when you don't want it to! If wine or beer is left open to the air for some time it oxidises and turns into vinegar, which is ethanoic acid.

Ethanoic acid is just like any other acid. It reacts with

- metals to form a salt + hydrogen
- bases and alkalis to form a salt + water
- carbonates to form a salt + water + carbon dioxide.

In each case the salt is called an **ethanoate**.

Esters

Organic acids like ethanoic acid react with alcohols to form compounds called **esters**. Many esters have strong smells. Flower scents are esters. Soap is made with esters too.

Soap

Beef fat contains an ester called **glyceryl stearate**. When the ester is boiled with steam and sodium hydroxide, hydrolysis takes place and the ester is split up into its acid **stearic acid** and alcohol, **glycerol**. The acid reacts further to become the salt **sodium stearate**.

glyceryl stearate + sodium hydroxide ⟶ sodium stearate + glycerol
    (beef fat)   (soap)

Sodium stearate is soap. Another name for soap making in this way is **saponification**. (See section 7.6 for more about soap.)

Detergents

Soap has one disadvantage. It is precipitated as **scum** when calcium and magnesium ions are in the water. Water of this sort is said to be hard water. (See section 7.8.) Synthetic detergents do the same washing job as soap, but are not affected by the hardness of the water. Synthetic **detergents**, sometimes called soapless detergents are manufactured from petrochemicals and sulphuric acid.

All in the packet

Persil contains the following:

- an alkylbenzene sulphonate **detergent** to remove the dirt.
- sodium tripolyphosphate, to **soften** the water by removing calcium and magnesium ions from the water.
- sodium silicate which acts as a **filler**, to make the powder flow smoothly and to reduce acidity in the washing water.
- sodium borate which is a **bleach** to remove stains.

- ethanolamide to **stabilise** the lather during washing, stopping it from becoming too frothy.
- sodium carboxymethyl cellulose to **stop dirt clinging** to cottons.
- a fluorescer to make the washing **look brighter**.
- a perfume to make washing **smell good**.

Did you know that Persil first came onto the market in 1909 and that its name comes from two chemicals in the powder called sodium **per**borate and sodium **sil**icate.

Figure 5
Persil was one of the first soapless detergents to be made in this country.

Key points

- Ethanol is made by fermentation.
- Fermented solutions can be concentrated by fractional distillation.
- Drinking alcohol can be dangerous.
- Acids like ethanoic acid form esters with alcohols.
- Soap is made when certain esters are hydrolysed with sodium hydroxide.
- Synthetic detergents are not affected by hardness of water, unlike soap, which is.

Quick questions

1 When a fermented liquid is fractionally distilled, the distillate contains almost pure ethanol. Describe a test you would do to show that this is true.

2 What chemical reaction would take place if (a) magnesium (b) copper oxide (c) sodium carbonate are added to ethanoic acid. In each case, describe what you would see, write the equation and name the products.

9.4 *Chemistry and food*

Pruteen

Humans and animals need protein in their diet. We get our protein from bread, grain, nuts, and by eating other animals like pigs, cows, chickens and sheep. These animals eat protein first, and then we eat them and so their protein gets to us second hand. In many ways, therefore, eating animals and not plants is a wasteful and expensive way of eating.

Since 1960, ICI has been growing protein, not on land or in the sea, but in the laboratory. The technicians use a bacterium that feeds on methanol, and they extract the protein material from its cells as it grows. Methanol is made from natural gas. Like all living things, the bacterium needs food and oxygen, so the cells of this microbe are mixed with water and methanol, and air and ammonia are bubbled through this suspension in a large container called a **fermenter** at a carefully controlled temperature of 33°C. (Figure 1.)

The bacterium cells grow very fast. At regular intervals they are removed and dried into granules. The final product is called **Pruteen** and contains 72% protein. It is used for feeding animals such as pigs, chicken, calves and turkeys.

Figure 1
This vast industrial complex produces Pruteen for livestock.

The world's food

In the Western World, 2% of the population grows food for the other 98%. If the food is fresh, like bread, fruit and vegetables, it must be delivered and used quickly. If the food is to be kept in stock until it is needed, it must be preserved. The western diet is usually complete in that it contains all the chemicals that we need, and we usually eat too much!

Whilst European countries in the EC have excess food many millions of people in the Third World exist on poor quality wheat, maize, rice, yams and cassava roots for their protein and carbohydrate, and they do not have enough to eat.

Protein is needed for growth and to make new cells to repair damaged tissue. Without it, people quickly grow listless and die. In famine areas, many children start to suffer from protein deficiency the moment they stop being breast-fed. If people do not get enough to eat, they also suffer from mineral and vitamin deficiency.

Although western countries produce 10% more food than they need, the world food problem is not just solved by giving away food although it is a last resort in times of famine. Instead, problems of over-population, education and climate must be considered too. What will happen to local rice farmers if a free supply of rice comes through aid, do you think?

Chemicals in food

Human food consists of different sorts of molecules, all of which are needed in different amounts to keep us healthy.

Protein is necessary to make new cells for growth and to repair worn out tissue. We get protein from meat, fish, cheese and eggs, as well as milk, bread and beans.

Carbohydrates provide energy and heat when they are oxidised in cells in our bodies. We get most of our carbohydrate from bread, potatoes, fruit and sugar. The amount of energy they give us is measured in kilojoules (kJ). The more energetic or hardworking you are, the more carbohydrate you need. A manual worker on a building site needs up to 20 000 kJ of energy from food each day.

Fats are necessary too, but in small quantities. Some of these are oxidised for energy but they may be stored as emergency supplies in the body. Fats come from dairy products such as butter (and margarine) and cheese as well as meat and fish.

If you eat more food than you need, the extra food is stored as fat layers under the skin.

Healthy fats

Some fat is necessary in a healthy diet, but some fats are better than others. Butter and fats from such things as lamb and pork contain **saturated fats**. They are called saturated because the carbon chains in the molecules of these compounds have no double bonds in them. Too much saturated fat allows **cholesterol** to be formed in the bloodstream, and heart disease can follow.

On the other hand, oils like olive oil and sunflower oil, and margarine made from these substances contain **polyunsaturated fats**. These molecules have several double bonds in them. They do not increase cholesterol and in some cases even prevent it forming. For this reason polyunsaturated fats are regarded as more healthy fats.

Vitamins

Different vitamins have different jobs in the body but they are all essential for health.

Vitamin A, found in fish oils, liver, green vegetables and milk, is needed for healthy skin, bones and eyes.

Vitamin B, found in meat, wholemeal bread, yeast, eggs, vegetables and fish, helps in the formation of blood and assists digestion.

Vitamin C, found in citrus fruits and green vegetables, helps wounds heal and keeps skin and gums in good condition.

Vitamin D, found in liver, eggs, fish and dairy products, and formed when the sun shines on your skin, is needed for bones and teeth.

Minerals

Minerals are compounds containing metals and are also important in a healthy diet. Muscles need sodium and potassium compounds. Red blood cells need iron and bones and teeth need calcium and phosphorus. Small amounts of these come in all sort of foods.

Other essentials

Water and fibre are also essential for a healthy diet. 98% of the human body **is** water! Fibre is material which cannot be digested and which passes through the digestive system unaffected. It is needed to give bulk to waste materials as they pass out of the intestines and into the bowel. Food scientists say that for a more healthy diet we should eat

> **less** saturated fat, white sugar, refined foods and salt
> **more** fibre, unrefined carbohydrates and polyunsaturated fat.

Food additives

If we were to be able to eat all food as soon as it was harvested from the ground or the moment it came out of the oven, food additives would not be necessary. Much of our food has to be stored and preserved so that we can eat it later in the year. Some additives are put into food to preserve it; others are used to make it look and taste good.

Figure 2
*Is this an essential part of
the child's diet?*

Preservatives prevent food from going bad by killing bacteria that might decompose it. Commonly used preservatives are sulphur dioxide and sodium sulphite for fruit; potassium nitrite for meat; and ethanoic acid (vinegar) for pickles. Salt and sugar are also used as preservatives.

Emulsifiers and stabilisers keep oils mixed up and bind foods together. These include polyphosphates and glyceryl esters. Alginates from seaweed and natural gums thicken and gel foods too.

Some additives make food **sweet** (sorbitol is often used instead of sucrose and glucose); **enhance flavour** (look for monosodium glutamate on the sides of packets); and give food a smell.

Food looks good if it is **coloured**. As well as natural colouring agents like cochineal, chlorophyll and tumeric, many colouring agents are made from coal and oil chemicals. In recent years it has been found that some colouring agents such as tartrazine and Sunset Yellow are responsible for skin disorders, asthmatic problems and hyperactivity in young children. Although widely used, for example in soft drinks, public complaints have led to these artificial colours being removed from many foods and being replaced by natural colours, or no colours at all.

E Numbers

By law, all packaged foods must give a lot of information on their labels. They must say who made them and where they were made too. The quantity of food must be stated and a list of ingredients must be given, in order of the amounts used. (Figure 4.)

Figure 3
By law, all tins and packets of food must give information about what is in the food. What additives are in this food?

Most additives are identified by an **E Number**. Here are some examples.

E number	Name	Use
E102	tartrazine	colouring
E110	Sunset Yellow	colouring
E150	caramel	colouring
E170	calcium carbonate	filler
E220	sulphur dioxide	preservative
E249	potassium nitrite	preservative
E270	lactic acid	preservative
E321	butylate hydroxytoluene	emulsifier
E440	pectin	setting agent
E621	monosodium glutamate	flavour enhancer

Key points

- Many people in the world do not have enough food to eat, or have food lacking in essential nutrients.
- A good diet contains sufficient protein, carbohydrate, fat, vitamins, minerals and fibre.
- Additives are put into food to preserve it or make it attractive.
- Most additives have an E number.

Quick questions

1 Why is carbohydrate necessary in our diet? Why is too much a bad thing?

2 Make a list of the additives in (a) an instant pudding (b) your local shop's own brand orange squash (c) packet soup (d) Smarties (e) flavoured crisps. In each case, look on the packet.

Questions

1 (a) In industrialised countries the atmosphere can become polluted due to the burning of fossil fuels.
    (i) What is meant by *fossil fuel*?
    (ii) Name three fossil fuels.

(b) Sulphur dioxide is formed during the burning of fossil fuels and is an atmospheric pollutant. The data below shows the average monthly concentrations in parts per million (ppm) of sulphur dioxide present in the air of a city centre.

Month	Jan	Feb	Mar	Apr	May	Jun	Jul	Aug	Sep	Oct	Nov	Dec
Concentration of SO_2/ppm	0.10	0.08	0.07	0.05	0.03	0.02	0.01	0.02	0.02	0.04	0.08	0.09

    (i) Represent this data in a suitable form on a graph using the scale shown below.

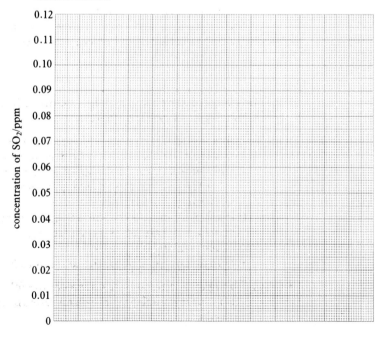

(ii) Why is sulphur dioxide pollution greatest during the winter months?

(c) Sulphur dioxide dissolves in rain water and slowly reacts with oxygen to form sulphuric acid.
   (i) Write a balanced equation to show how sulphuric acid is formed in this process.
   (ii) Describe a simple test that can be used to measure the acidity of rain water.
   (iii) Give one reason why 'acid rain' is considered a serious nuisance.

SEG

⋆ **2 (a)** Name **and** give the formula of the main compound in natural gas.

(b) A compound found in the oil fraction kerosene is decane, $C_{10}H_{22}$.
   (i) Complete the word equation for burning decane completely in air.

   decane + oxygen $\longrightarrow$

   (ii) What conditions might cause a poisonous gas to be formed when burning decane?
   (iii) Name the poisonous gas referred to in **(b)**(ii)

(c) Polythene [poly(ethene)] is made by joining together many ethene molecules.
   (i) What is this process called?
   (ii) The formula of polythene is

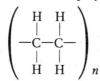

   What does the 'n' mean?
   (iii) Draw the structure (structural formula) of ethene.
   (iv) What feature of the ethene molecule allows it to be changed into polythene.

SEG

3 The general formula for the alkene series of hydrocarbons is C_nH_{2n}. Ethene, C_2H_4, is the first member.

(a) (i) Work out the molecular formula of the fifth member, hexene.
   (ii) Ethene may be prepared by passing ethanol vapour over a heated aluminium oxide catalyst. Sketch an apparatus which you might use to carry out this reaction, showing how you would collect the ethene.
   (iii) What would you do when using the apparatus shown in your answer to part (ii) to make sure that the ethene was reasonably free of air?
   (iv) Write an equation for the reaction in (ii).

⋆ **(b)** An organic compound X undergoes the following reactions.
   (i) It burns completely in oxygen forming carbon dioxide and water only.

(ii) It rapidly decolourizes bromine water.

(iii) It dissolves in sodium carbonate solution with fizzing.

State as fully as possible what you can deduce about the structure of compound X from each reaction.

★ **(c)** Propene has the structural formula shown below.

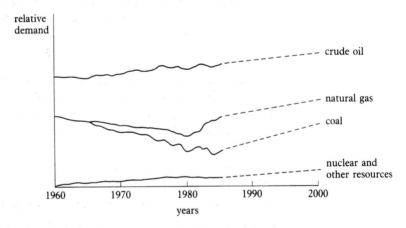

Write down the structural formulae for the products of the following reactions of propene:

(i) hydrogenation;

(ii) hydration;

(iii) polymerisation.

(d)   (i) State one environmental disadvantage of a polymer such as poly(propene).

(ii) Explain why poly(propene) is much less reactive than propene.

SEG

4 The chart shows the demand for the types of energy resources.

relative demand

crude oil

natural gas

coal

nuclear and other resources

1960   1970   1980   1990   2000

years

(a) Name two resources that between 1960 and 1965 were used up to the same extent.

(b) Which named resource is now in less demand that it was in 1960?

(c) Name two other energy resources not named in the chart.

(d) Crude oil is a finite resource. Explain the meaning of the words *finite* and *resource*.

(e) Why are the energy demand lines dotted after 1986?

(f) Wet decaying plant or animal material ferments in the absence of air to form biogas. Biogas is mainly a mixture of methane (CH_4) and carbon dioxide (CO_2).

(i) Why is biogas classed as a fuel?

(ii) If you had a biogas generator for your home what type of household rubbish could it use?

(g) If biogas is passed through potassium hydroxide solution all the carbon dioxide is removed. The percentage of carbon dioxide in biogas can be found by measuring the volume of gas remaining.

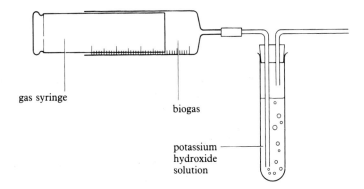

gas syringe

biogas

potassium hydroxide solution

(i) Copy and complete the diagram above which shows part of the apparatus for carrying out this experiment.

(ii) When 90cm^3 of biogas was used, the volume of gas remaining after passing through the potassium hydroxide solution was 36 cm^3. Calculate the precentage by volume of carbon dioxide in the biogas.

SEG

5 This question is about the formation of alcohol (ethanol) from sugars and its possible use as an alternative to petrol as a fuel for car engines.

One source of sugars is sugar cane which is crushed and the juices mixed with yeast. The mixture is allowed to stand for two or three days at around 30°C.

The liquid product is then fractionally distilled, most of the ethanol being in the middle of three fractions.

(a) Name **one** other crop which is a useful source of sugars.

(b) The equation for the reaction which changes the sugar glucose into ethanol in the presence of yeast is given below.

$$C_6H_{12}O_6(aq) \longrightarrow 2\,C_2H_5OH(aq) + 2\,CO_2(g)$$

(i) What does the symbol (aq) indicate about the glucose?

(ii) What is the purpose of the yeast in the reaction?

(iii) Why is this reaction **not** speeded up if the mixture is boiled?

(iv) Give the name of the process which converts glucose into ethanol in this way.

(v) Why is the same reaction important in bread-making?

(c) One of the advantages of ethanol over petrol is that, unlike petrol, ethanol is a *renewable energy source*.

Explain the meaning of the term *renewable energy source*.

(d) What other possible advantages might ethanol have over petrol as a fuel for car engines?

(e) Methylated spirits is a mixture of ethanol (about 90%) and methanol (about 10%) together with a small quantity of purple dye.

Explain why the ethanol is treated in this way before being sold as 'meths'.

SEG

THE PERIODIC TABLE

Groups

Noble gases

Group I	Group II											III	IV	V	VI	VII	He (4, 2)
H																	
Li (7, 3)	Be (9, 4)					transition metals						B (11, 5)	C (12, 6)	N (14, 7)	O (16, 8)	F (19, 9)	Ne (20, 10)
Na (23, 11)	Mg (24, 12)											Al (27, 13)	Si (28, 14)	P (31, 15)	S (32, 16)	Cl (35.5, 17)	Ar (40, 18)
K (39, 19)	Ca (40, 20)	Sc (45, 21)	Ti (48, 22)	V (51, 23)	Cr (52, 24)	Mn (55, 25)	Fe (56, 26)	Co (59, 27)	Ni (59, 28)	Cu (64, 29)	Zn (65, 30)	Ga (70, 31)	Ge (73, 32)	As (25, 33)	Se (79, 34)	Br (80, 35)	Kr (84, 36)
Rb (85.5, 37)	Sr (88, 38)	Y (89, 39)	Zr (91, 40)	Nb (93, 41)	Mo (96, 42)	Tc (98, 43)	Ru (101, 44)	Rh (103, 45)	Pd (106, 46)	Ag (108, 47)	Cd (112, 48)	In (115, 49)	Sn (119, 50)	Sb (122, 51)	Te (128, 52)	I (127, 53)	Xe (131, 54)
Cs (132, 55)	Ba (137, 56)	La (139, 57)	Hf (178.5, 72)	Ta (181, 73)	W (184, 74)	Re (186, 75)	Os (190, 76)	Ir (192, 77)	Pt (195, 78)	Au (197, 79)	Hg (201, 80)	Tl (204, 81)	Pb (207, 82)	Bi (209, 83)	Po (210, 84)	At (210, 85)	Rn (223, 86)
Fr (223, 87)	Ra (226, 88)	Ac (227, 89)	Unq (104)	Unp (105)	Unh (106)	Uns (107)	Uno (108)	Une (109)									

La (139, 57)	Ce (140, 58)	Pr (141, 59)	Nd (144, 60)	Pm (147, 61)	Sm (150, 62)	Eu (152, 63)	Gd (157, 64)	Tb (159, 65)	Dy (162.5, 66)	Ho (165, 67)	Er (167, 68)	Tm (169, 69)	Yb (173, 70)	Lu (175, 71)
Ac (227, 89)	Th (232, 90)	Pa (231, 91)	U (238, 92)	Np (237, 93)	Pu (242, 94)	Am (243, 95)	Cm (247, 96)	Bk (247, 97)	Cf (251, 98)	Es (254, 99)	Fm (253, 100)	Md (256, 101)	No (254, 102)	Lw (257, 103)

Mass Number

SYMBOL

Atomic Number

Elements with atomic numbers of more than 100 are now named by letters representing their numbers:

0 = nil   2 = bi   4 = quad   6 = hex   8 = oct
1 = un   3 = tri   5 = pent   7 = sept   9 = enn

So element 110 will be **unun** and it will be called **ununnilium**

Approximate relative atomic masses (for calculations)

Aluminium	Al	27
Barium	Ba	127
Bromine	Br	80
Calcium	Ca	40
Carbon	C	12
Chlorine	Cl	35.5
Copper	Cu	64
Hydrogen	H	1
Iodine	I	127
Iron	Fe	56
Lead	Pb	207
Lithium	Li	7
Magnesium	Mg	24
Manganese	Mn	55
Mercury	Hg	201
Nitrogen	N	14
Oxygen	O	16
Phosphorus	P	31
Potassium	K	39
Silicon	Si	28
Silver	Ag	108
Sodium	Na	23
Sulphur	S	32
Zinc	Zn	65

Relative atomic masses based on internationally agreed figures.

Element	Symbol	Atomic number	Relative atomic mass	Element	Symbol	Atomic number	Relative atomic mass
Actinium	Ac	89		Mercury	Hg	80	200·59
Aluminium	Al	13	26·9815	Molybdenum	Mo	42	95·94
Americium	Am	95		Neodymium	Nd	60	144·24
Antimony	Sb	51	121·75	Neon	Ne	10	20·179
Argon	Ar	18	39·948	Neptunium	Np	93	
Arsenic	As	33	74·9216	Nickel	Ni	28	58·71
Astatine	At	85		Niobium	Nb	41	92·906
Barium	Ba	56	137·34	Nitrogen	N	7	14·0097
Berkelium	Bk	97		Nobelium	No	102	
Beryllium	Be	4	9·0122	Osmium	Os	76	190·2
Bismuth	Bi	83	208·980	Oxygen	O	8	15·9994
Boron	B	5	10·811	Palladium	Pd	46	106·4
Bromine	Br	35	79·909	Phosphorus	P	15	30·9738
Cadmium	Cd	48	112·40	Platinum	Pt	78	195·09
Caesium	Cs	55	132·905	Plutonium	Pu	94	
Calcium	Ca	20	40·08	Polonium	Po	84	
Californium	Cf	98		Potassium	K	19	39·102
Carbon	C	6	12·01115	Praseodymium	Pr	59	140·907
Cerium	Ce	58	140·12	Promethium	Pm	61	
Chlorine	Cl	17	35·453	Protactinium	Pa	91	
Chromium	Cr	24	51·996	Radium	Ra	88	
Cobalt	Co	27	58·9332	Radon	Rn	86	
Copper	Cu	29	63·54	Rhenium	Re	75	186·2
Curium	Cm	96		Rhodium	Rh	45	102·905
Dysprosium	Dy	66	162·50	Rubidium	Rb	37	85·47
Einsteinium	Es	99		Ruthenium	Ru	44	101·07
Erbium	Er	68	167·26	Samarium	Sm	62	150·35
Europium	Eu	63	151·96	Scandium	Sc	21	44·956
Fermium	Fm	100		Selenium	Se	34	78·96
Fluorine	F	9	18·9984	Silicon	Si	14	28·086
Francium	Fr	87		Silver	Ag	47	107·868
Gadolinium	Gd	64	157·25	Sodium	Na	11	22·9898
Gallium	Ga	31	69·72	Strontium	Sr	38	87·62
Germanium	Ge	32	72·59	Sulphur	S	16	32·064
Gold	Au	79	196·967	Tantalum	Ta	73	180·948
Hafnium	Hf	72	178·49	Technetium	Tc	43	
Helium	He	2	4·0026	Tellurium	TE	52	127·60
Holmium	Ho	67	164·930	Terbium	Tb	65	158·924
Hydrogen	H	1	1·00797	Thallium	Tl	81	204·37
Indium	In	49	114·82	Thorium	Th	90	232·038
Iodine	I	53	126·9044	Thulium	Tm	69	168·934
Iridium	Ir	77	192·2	Tin	Sn	50	118·69
Iron	Fe	26	55·847	Titanium	Ti	22	47·90
Krypton	Kr	36	83·80	Tungsten	W	74	183·85
Lanthanum	La	57	138·91	Uranium	U	92	238·03
Lawrencium	Lw	103		Vanadium	V	23	50·942
Lead	Pb	82	207·19	Xenon	Xe	54	131·30
Lithium	Li	3	6·939	Ytterbium	Yb	70	173·04
Lutetium	Lu	71	174·97	Yttrium	Y	39	88·905
Magnesium	Mg	12	24·312	Zinc	Zn	30	65·37
Manganese	Mn	25	54·9380	Zirconium	Zr	40	91·22
Mendelevium	Md	101					

Index